European Studies in Philosophy of Science

Volume 6

This new series results from the synergy of EPSA - European Philosophy of Science Association - and PSE - Philosophy of Science in a European Perspective: ESF Networking Programme (2008–2013). It continues the aims of the Springer series "The Philosophy of Science in a European Perspective" and is meant to give a new impetus to European research in the philosophy of science. The main purpose of the series is to provide a publication platform to young researchers working in Europe, who will thus be encouraged to publish in English and make their work internationally known and available. In addition, the series will host the EPSA conference proceedings, selected papers coming from workshops, edited volumes on specific issues in the philosophy of science, monographs and outstanding Ph.D. dissertations. There will be a special emphasis on philosophy of science originating from Europe. In all cases there will be a commitment to high standards of quality. The Editors will be assisted by an Editorial Board of renowned scholars, who will advise on the selection of manuscripts to be considered for publication.

More information about this series at http://www.springer.com/series/13909

Gábor Hofer-Szabó • Leszek Wroński

Editors

Making it Formally Explicit

Probability, Causality and Indeterminism

 Springer

Editors
Gábor Hofer-Szabó
Institute of Philosophy
Research Centre for the Humanities
Hungarian Academy of Sciences
Budapest, Hungary

Leszek Wroński
Jagiellonian University
Kraków, Poland

ISSN 2365-4228 ISSN 2365-4236 (electronic)
European Studies in Philosophy of Science
ISBN 978-3-319-55485-3 ISBN 978-3-319-55486-0 (eBook)
DOI 10.1007/978-3-319-55486-0

Library of Congress Control Number: 2017937660

Printed on acid-free paper

This Springer imprint is published by Springer Nature
The registered company is Springer International Publishing AG
The registered company address is: Gewerbestrasse 11, 6330 Cham, Switzerland

Preface

The present book collects research papers related to talks presented at six workshops which were organized by *The Budapest-Kraków Research Group on Probability, Causality and Determinism* and held alternatingly in Budapest and Kraków during 2014–2016. The members of the group are grateful to the speakers and attendees of these workshops for the valuable discussions from which the present papers greatly benefited. We also acknowledge the generous support of the Hungarian Scientific Research Fund, OTKA K-115593, and the Bilateral Mobility Grant of the Hungarian and Polish Academies of Sciences, NM-104/2014.

Budapest, Hungary
Kraków, Poland

Gábor Hofer-Szabó
Leszek Wroński

Contents

**Part III Indeterminism, Undecidability,
            and Macrostates**

About the Authors

Gergei Bana is a research engineer at INRIA, the French Institute for Research in Computer Science and Automation. His primary research activity concerns automated verification of cryptographic protocols through first-order logical representations of complexity theoretic properties. He has also been interested in foundational questions of quantum theory and probability theory and recently has been working on problems related to Lewis's Principal Principle.

Juliusz Doboszewski is a PhD student at the Institute of Philosophy of Jagiellonian University in Kraków. He is working on the foundations of classical general relativity. He has also side interests in the philosophy of probability and mathematical logic.

László E. Szabó is a professor in the Department of Logic, Institute of Philosophy, at Eötvös Loránd University, Budapest. His research focuses on the philosophy of space and time, causality, the EPR-Bell problem, the interpretation of probability, and a physicalist account of mathematics.

Samuel C. Fletcher is an assistant professor of philosophy at the University of Minnesota, Twin Cities, and Marie Curie Fellow at the Munich Center for Mathematical Philosophy, Ludwig Maximilian University. Much of his work so far has concerned the foundations of physics and of statistics and how problems in these fields inform and are informed by broader issues in the philosophy of science. He also has interests in the conceptual and physical basis of computation, the philosophy of applied mathematics, and the history of physics and philosophy of science.

Michał Tomasz Godziszewski is a PhD student in the Logic Department of the Institute of Philosophy at the University of Warsaw. For his thesis he is working on the model theory of arithmetic and formal theories of truth. He has graduated from the Department of Interfaculty Studies in Humanities at the University of Warsaw (studying mainly philosophy and mathematics). His scientific interests concern logic and its applications, foundations of mathematics, philosophy of science, and formal epistemology.

Márton Gömöri is a junior research fellow in the Institute of Philosophy, Research Centre for the Humanities, at the Hungarian Academy of Sciences. His main fields of interest are foundations of relativity theory and probability in physics.

Balázs Gyenis is a research fellow in the Institute of Philosophy, Research Centre for the Humanities, at the Hungarian Academy of Sciences who recently received his PhD from the Department of History and Philosophy of Science at the University of Pittsburgh. His main fields are philosophy of physics, general philosophy of science, determinism, and causality.

Zalán Gyenis is an MTA Postdoctoral Research Fellow in the Department of Algebra at the Budapest University of Technology and Economics. His main fields are logic and model theory.

Gábor Hofer-Szabó is a senior research fellow in the Institute of Philosophy, Research Centre for the Humanities, at the Hungarian Academy of Sciences. His main fields of research are foundations of quantum mechanics, interpretations of probability, and probabilistic causality.

Yuichiro Kitajima is an assistant professor in the College of Industrial Technology at Nihon University, Japan. His current research focuses on foundational and philosophical problems of quantum mechanics. In particular, he deals with reality and causality in algebraic quantum field theory. Some topics of his papers are Reichenbachian common causes, Bell inequalities, and Bohr's reply to EPR.

Tomasz Placek is a professor at Jagiellonian University in Kraków. His current field of interest is metaphysics and philosophy of physics.

Miklós Rédei is the professor and head in the Department of Philosophy, Logic and Methodology of Science at the London School of Economics and Political Science. His research interests are philosophy and the foundations of physics.

Iñaki San Pedro has held research positions at the Complutense University of Madrid, CPNSS-London School of Economics, and UPV/EHU. His main research interests include causal inference in genuinely indeterministic contexts and the philosophical foundations of quantum mechanics.

Christian Wallmann is a postdoctoral researcher in Jon Williamson's project on "Evaluating Evidence in Medicine." He holds a PhD in philosophy from the University of Salzburg and a BSc in mathematics from Ludwig Maximilian University of Munich. Christian's research interests are in the philosophy of science, philosophy of medicine, formal epistemology, and probabilistic logic. His current research focuses on the reference class problem, diagnosis, and prediction and on evidence of mechanism in medicine.

Jon Williamson works on the philosophy of causality, the foundations of probability, formal epistemology, inductive logic, and the use of causality, probability, and inference methods in science and medicine. He is the professor of reasoning, inference, and scientific method in the Philosophy Department at the University of Kent. He is a member of the Theoretical Reasoning research cluster there and a codirector of the Centre for Reasoning. He is the editor of The Reasoner.

Leszek Wroński is an assistant professor at the Institute of Philosophy of Jagiellonian University in Kraków. He is mainly interested in formal epistemology and philosophy of probability, having authored and coauthored also papers on probabilistic causality and the branching space-time theory.

Introduction

This book contains research articles on the philosophical foundations of probability, causality, space-time, and quantum theory. It offers a collection of papers in philosophy of science practiced in a formally rigorous and conceptually precise style characteristic to *The Budapest-Kraków Research Group*. This research group was established in 2014 with the aim of engaging in discussion, readings, workshops, and research projects on issues related to probabilistic causality, foundations of physics, and Bayesian epistemology. The group is led by Gábor Hofer-Szabó (Budapest) and Leszek Wroński (Kraków); the members are Juliusz Doboszewski, Márton Gömöri, Balázs Gyenis, Zalán Gyenis, and Tomasz Placek. The work of the group is often characterized by the emphasis put on the importance of making explicit the mathematical-structural assumptions that underlie philosophical argumentations. The group members are convinced that paying insufficient attention to the finer formal details often aggravates the philosophical problems that are in the center of focus of contemporary philosophical research. The group organizes two workshops in each year: one in Budapest and one in Kraków. For the present book, we have chosen 13 papers related to some of the roughly 70 talks of six subsequent workshops.

The book consists of three parts, connected to the three themes in the title of the book: probability, causality, and indeterminism. Part I focuses on the notion of probability from a general philosophical and formal epistemological perspective. Part II applies probabilistic considerations to causal questions in the foundations of quantum mechanics. Finally, Part III investigates the question of indeterminism in space-time theories and some related questions such as physical theory, decidability, and observation.

Part I concerns probability and chance-credence norms. The first three papers all deal with different aspects of the Principal Principle introduced by David Lewis. First, Miklós Rédei and Zalán Gyenis introduce the issue of proper mathematical formulation of the principle and tackle the problem of its consistency. Next, Balázs Gyenis and Leszek Wroński study a recent proposal which links the Principal Principle to the Principle of Indifference, obtaining some generalized results. Following that, Gergei Bana investigates how objective chance can be incorporated in the subjectivist's model in a mathematically rigorous manner. In the last paper of Part I, Christian Wallmann and Jon Williamson present and analyze four approaches to the reference class problem.

The topic of Part II is that of various probabilistic structures used in modeling quantum (but not only quantum) experiments. First, Leszek Wroński, Tomasz Placek, and Michał T. Godziszewski present new results regarding the possibility of explaining the Bell correlations by means of "separate common cause systems." In the next paper, Márton Gömöri and Tomasz Placek state and analyze Bell's theorem in terms of the so-called "small probability space" formalism. Then, Yuichiro Kitajima presents a new result concerning Reichenbachian common cause systems in general probability theories, which include both classical and quantum theories. Lastly, Iñaki San Pedro discusses the rejection of the measurement independence (no-conspiracy) condition in the context of causal explanations of EPR correlations.

The topic of Part III is physical theory, indeterminism, and macrostates. In the first paper László E. Szabó sketches a coherent account for physical theory, within the philosophical framework spanned by three doctrines: physicalism, empiricism, and the formalist philosophy of mathematics. Next, Samuel C. Fletcher investigates the so-called Lipschitz indeterminism of space-time geodesics and its implications for indeterministic space-time models. Following that, Juliusz Doboszewski studies space-time extensions with the aim of making the notion of indeterminism more precise. Then, Márton Gömöri, Balázs Gyenis, and Gábor Hofer-Szabó present a new idea on how macrostates come about in statistical mechanics as partitions of the phase space due to the interaction between the system and the observer. The last paper of the book is by Michał Tomasz Godziszewski, who analyzes the question concerning the existence of a reliable computational procedure for finding certain causal and deterministic relationships in physical theories.

We hope that such an anthology of papers may help to increase the visibility of the distinct philosophical flavor that arises from an emphasis on making philosophical questions and assumptions formally explicit.

Part I
Probability and Chance-Credence Norms

Chapter 1
A Principled Analysis of Consistency of an Abstract Principal Principle

Zalán Gyenis and Miklós Rédei

Abstract The paper takes the Abstract Principal Principle to be a norm demanding that subjective degrees of belief of a Bayesian agent be equal to the objective probabilities once the agent has conditionalized his subjective degrees of beliefs on the values of the objective probabilities, where the objective probabilities can be not only chances but any other quantities determined objectively. Weak and strong consistency of the Abstract Principal Principle are defined in terms of both classical an non-classical (quantum) probability measure spaces. It is proved that the Abstract Principal Principle is weakly and strongly consistent both in the classical and non-classical cases. It is argued that it is desirable to strengthen the Abstract Principal Principle by adding a stability requirement to it. Weak and strong consistency of the resulting Stable Abstract Principal Principle are defined, and the strong consistency of the Abstract Principal Principle is interpreted as necessary for a non-omniscient Bayesian agent to be able to have rational degrees of belief in all epistemic situations. We give a proof of weak consistency of the Stable Abstract Principal Principle in the framework of classical probability theory, a proof which is different from the proof in Bana (Philos Sci, 2016, forthcoming). Formulation and investigation of the status of a Stable Abstract Principal Principle in quantum probability spaces remains for further study.

1.1 The Claims

The aim of this paper is to investigate the consistency of what we call the "Abstract Principal Principle". The analysis carried out here elaborates in more detail and develops further the measure theoretic analysis of the Principal Principle initiated in

Z. Gyenis (✉)
Department of Algebra, Budapest University of Technology and Economics, Budapest, Hungary
e-mail: gyz@renyi.hu

M. Rédei
Department of Philosophy, Logic and Scientific Method, London School of Economics and Political Science, Houghton Street, London WC2A 2AE, UK
e-mail: m.redei@lse.ac.uk

© Springer International Publishing AG 2017
G. Hofer-Szabó, L. Wroński (eds.), *Making it Formally Explicit*, European Studies in Philosophy of Science 6, DOI 10.1007/978-3-319-55486-0_1

Gyenis and Rédei (2016): Here we extend the analysis to non-classical (quantum) probability spaces.

We take the Abstract Principal Principle to be a general norm that regulates probabilities representing the subjective degrees of belief of an abstract Bayesian agent by requiring the agent's degrees of belief to be equal to the objective probabilities if the agent knows the values of the objective probabilities. We call this principle the *Abstract* Principal Principle because nothing is assumed about the specific nature of the objective probabilities – they can be (finite or infinite) relative frequencies, chances, propensities, ratios of some sort, or any other quantities viewed as determined objectively, i.e. independently of the agent and his beliefs.

After stating the Abstract Principal Principle informally in Sect. 1.2 we describe in a non-technical way the consistency problem to be analyzed in the paper. The consistency in question is of a fundamental nature: it expresses the harmony of the Abstract Principal Principle with the basic structure of measure theoretic probability theory. It will be seen that this consistency comes in different degrees of strength and we develop them step by step, proceeding from weaker to stronger. In Sect. 1.3 we define formally, in terms of classical measure theoretic probability theory specified by the standard Kolmogorovian axioms, the *weak consistency* of the Abstract Principal Principle (Definition 3.1) and prove that the Abstract Principal Principle is weakly consistent (Proposition 3.2). The proof will reveal a weakness in the concept of weak consistency and in Sect. 1.4 we will define the *strong consistency* of the Abstract Principal Principle. We then prove (details of the proof are given in the Appendix) that, under some (mild) assumptions on the agent's prior subjective probability, the Abstract Principal Principle is strongly consistent (Proposition 4.4). We will then argue that it is very natural to strengthen the Abstract Principal Principle by requiring it to satisfy a *stability* property, which expresses that conditional degrees of belief in events *already* equal (in the spirit of the Abstract Principal Principle) to the objective probabilities of the events do not change as a result of conditionalizing them further on knowing the objective probabilities of *other* events (in particular of events that are independent with respect to their objective probabilities). We call this amended principle the *Stable* Abstract Principal Principle (if stability is required only with respect to further conditionalizing on values of probabilities of *independent* events: *Independence-Stable* Principal Principle). This stability requirement leads to suitably modified versions of both the weak and strong consistency of the (*Independence-*)Stable Abstract Principal Principle (Definitions 6.1 and 6.4). We will prove that the Stable Abstract Principal Principle is *weakly* consistent (Proposition 6.2), irrespective of cardinality of the Boolean algebra of random events. This entails that the Independence-Stable Abstract Principal Principle also is weakly consistent (Proposition 6.3). (Details of the proof are given in the Appendix.)

The *strong* consistency of both the Stable and the Independence-Stable Abstract Principal Principle was proved by Bana (2016) using techniques different from the ones in this paper, and Bana's interpretation of the Principal Principle also is somewhat different from the position we take in this paper. Section 1.7 will discuss the conceptual significance of the strong consistency of the Stable Abstract Principal

Principle by arguing that it is a necessary condition for a non-omniscient Bayesian agent to be able to have rational degrees of belief under all epistemic conditions.

Throughout the systematic part of the paper containing the results (Sects. 1.2, 1.3, 1.4, 1.5, 1.6, and 1.7) no references will be given to relevant and related literature. Section 1.8 puts the results into the context of the rich and growing literature on the Principal Principle. In particular, we discuss in this section the relevance of the notion of strong consistency of the Stable Abstract Principal Principle from the perspective of the Principal Principle about chances. The main message of this section is that the strong consistency of the Stable Abstract Principal Principle also is necessary for the consistency of both the original formulation of the Principal Principle by David Lewis and for the consistency of some of the subsequent modifications of Lewis' Principal Principle that have been proposed in the literature. We will also argue in this section that one would need consistency proofs similar to the one discussed in this paper concerning the suggested modifications of Lewis' Principal Principle in order to show that those modifications also are in harmony with the basic conceptual structure of measure theoretic probability. To the best of our knowledge no such proofs have been given yet.

In the final Sect. 1.9 we turn our attention to non-classical probability spaces. A non-classical probability space is the projection lattice of a von Neumann algebra acting on a separable Hilbert space together with a normal state playing the role of the classical probability measure. Section 1.9 formulates the problem of consistency of the Abstract Principal Principle in the context of such non-classical probability spaces. We call the corresponding principle the General Principal Principle. Theorem 9.2 (the proof of which can be found in the Appendix) proves that the General Principal Principle is strongly consistent.

1.2   The Abstract Principal Principle Informally

The Abstract Principal Principle regulates probabilities representing the subjective degrees of belief of an abstract Bayesian agent by stipulating that the subjective degrees of belief $p_{subj}(A)$ of the agent in events A are related to the objective probabilities $p_{obj}(A)$ as

$$p_{subj}(A|^\ulcorner p_{obj}(A) = r^\urcorner) = r \tag{1.1}$$

where $^\ulcorner p_{obj}(A) = r^\urcorner$ denotes the proposition "the objective probability, $p_{obj}(A)$, of A is equal to r".

The Abstract Principal Principle – and the formulation given by Eq. (1.1) in particular – presupposes that both p_{subj} and p_{obj} have the features of a probability measure: they both are assumed to be additive maps defined on a Boolean algebra taking values in the unit interval $[0, 1]$: p_{obj} is supposed to be defined on a Boolean algebra $\mathcal{S}_{obj}$ of random events viewed as specified objectively (equivalently, on a Boolean algebra of propositions stating that the random events happen); and p_{subj} also is supposed to be a map with a domain of definition being a Boolean algebra $\mathcal{S}_{subj}$.

The starting point of the analysis is the observation Gyenis and Rédei (2016) that for the conditional probability $p_{subj}(A|\ulcorner p_{obj}(A) = r\urcorner)$ in Eq. (1.1) to be well-defined via Bayes' rule, it is necessary that the Boolean algebra $\mathcal{S}_{subj}$ serving as the domain of definition of the probability measure p_{subj} contains *both* the Boolean algebra $\mathcal{S}_{obj}$ of random events *and* with every random event A also the proposition $\ulcorner p_{obj}(A) = r\urcorner$ – for if $\mathcal{S}_{subj}$ does not contain both of these two sorts of propositions then the formula $p_{subj}(A|\ulcorner p_{obj}(A) = r\urcorner)$ cannot be interpreted as an expression of conditional probability specified by Bayes' rule because the conditional probability $p_{subj}(A|\ulcorner p_{obj}(A) = r\urcorner)$ given by Bayes' rule reads

$$p_{subj}(A|\ulcorner p_{obj}(A) = r\urcorner) = \frac{p_{subj}(A \cap \ulcorner p_{obj}(A) = r\urcorner)}{p_{subj}(\ulcorner p_{obj}(A) = r\urcorner)} \tag{1.2}$$

thus all three propositions A, $\ulcorner p_{obj}(A) = r\urcorner$ and $A \cap \ulcorner p_{obj}(A) = r\urcorner$ must belong to the Boolean algebra $\mathcal{S}_{subj}$ on which the subjective probability p_{subj} is defined.

It is far from obvious however that, given *any* Boolean algebra $\mathcal{S}_{obj}$ of random events with *any* probability measure p_{obj} on $\mathcal{S}_{obj}$, there exists a Boolean algebra $\mathcal{S}_{subj}$ meeting these algebraic requirements in such a way that a probability measure p_{subj} satisfying the condition (1.2) also exists on $\mathcal{S}_{subj}$. If there exists a Boolean algebra $\mathcal{S}^*_{obj}$ of random events with a probability measure p^*_{obj} giving the objective probabilities of events for which there exists *no* Boolean algebra $\mathcal{S}_{subj}$ on which a probability function p_{subj} satisfying (1.2) can be defined, then the Abstract Principal Principle would not be maintainable in general – the Abstract Principal Principle would then be inconsistent as a general norm: In this case the agent, being in the epistemic situation of facing the objective facts represented by $(\mathcal{S}^*_{obj}, p^*_{obj})$, cannot have degrees of belief satisfying the Abstract Principal Principle for fundamental structural reasons inherent in the basic structure of classical probability theory. We say that the Abstract Principal Principle is *weakly consistent* if it is *not* inconsistent in the sense described. (The adjective "weakly" will be explained shortly.) To formulate the weak consistency of the Abstract Principal Principle precisely, we fix some notation and recall some definitions first.

1.3   Weak Consistency of the Abstract Principal Principle

The triplet $(X, \mathcal{S}, p)$ denotes a classical probability measure space specified by the Kolmogorovian axioms, where X is the set of elementary random events, $\mathcal{S}$ is a Boolean algebra of (some) subsets of X representing a general event and p is a probability measure on $\mathcal{S}$. Given two Boolean algebras $\mathcal{S}$ and $\mathcal{S}'$, the map $h: \mathcal{S} \to \mathcal{S}'$ is a Boolean algebra embedding if it preserves all Boolean operations:

$$h(A_1 \cup A_2) = h(A_1) \cup h(A_2) \tag{1.3}$$

$$h(A_1 \cap A_2) = h(A_1) \cap h(A_2) \qquad (1.4)$$

$$h(A^{\perp}) = h(A)^{\perp} \qquad (1.5)$$

and is injective:

$$A \neq B \text{ entails } h(A) \neq h(B) \qquad (1.6)$$

If $h: \mathcal{S} \to \mathcal{S}'$ is a Boolean algebra embedding, then $h(\mathcal{S})$ is an isomorphic copy of $\mathcal{S}$, which can be viewed as a Boolean subalgebra of $\mathcal{S}'$. From the perspective of probability theory elements A and $h(A)$ can be regarded as identical $A \leftrightarrow h(A)$.

The probability space $(X', \mathcal{S}', p')$ is called an extension of the probability space $(X, \mathcal{S}, p)$ with respect to h if h is a Boolean algebra embedding of $\mathcal{S}$ into $\mathcal{S}'$ that preserves the probability measure p:

$$p'(h(A)) = p(A) \qquad A \in \mathcal{S} \qquad (1.7)$$

Definition 3.1 The Abstract Principal Principle is defined to be *weakly consistent* if the following hold: Given any probability space $(X_{obj}, \mathcal{S}_{obj}, p_{obj})$, there exists a probability space $(X_{subj}, \mathcal{S}_{subj}, p_{subj})$ and a Boolean algebra embedding h of $\mathcal{S}_{obj}$ into $\mathcal{S}_{subj}$ such that

(i) For every $A \in \mathcal{S}_{obj}$ there exists an $A' \in \mathcal{S}_{subj}$ with the property

$$p_{subj}(h(A)|A') = p_{obj}(A) \qquad (1.8)$$

(ii) If $A, B \in \mathcal{S}_{obj}$ and $A \neq B$ then $A' \neq B'$.

The intuitive content of Definition 3.1 should now be clear: The probability space $(X_{obj}, \mathcal{S}_{obj}, p_{obj})$ describes the objective probabilities; in particular $p_{obj}(A)$ is the probability of the random event $A \in \mathcal{S}_{obj}$. The Boolean algebra $\mathcal{S}_{subj}$ in the probability space $(X_{subj}, \mathcal{S}_{subj}, p_{subj})$ contains not only the "copies" $h(A)$ of all the random events $A \in \mathcal{S}_{obj}$ (together with all the undistorted algebraic relations among the random events), but, with every random event $A \in \mathcal{S}_{obj}$, also an element A' to be interpreted as representing the proposition "the objective probability, $p_{obj}(A)$, of A is equal to r" (this proposition we denoted by $\ulcorner p_{obj}(A) = r \urcorner$). If $A \neq B$ then $A' \neq B'$ must be the case because $\ulcorner p_{obj}(A) = r \urcorner$ and $\ulcorner p_{obj}(B) = s \urcorner$ are different propositions – this is expressed by (ii) in the definition. The main content of the Abstract Principal Principle is then expressed by condition (1.8), which states that the *conditional* degrees of beliefs $p_{subj}(h(A)|A')$ of an agent about random events $h(A) \leftrightarrow A \in \mathcal{S}_{obj}$ are equal to the objective probabilities $p_{obj}(A)$ of the random events, where the condition A' is that the agent knows the values of the objective probabilities.

Proposition 3.2 (Gyenis and Rédei 2016) *The Abstract Principal Principle is weakly consistent.*

The above proposition follows from the proposition that states the weak consistency of the *Stable* Abstract Principal Principle (Proposition 6.2), which we state later and prove in the Appendix. To motivate the need to strengthen the notion of weak consistency, we give here a proof of Proposition 3.2 under the simplifying restriction that the Boolean algebra S_{obj} has a finite number of elements. The proof will expose the conceptual weakness of the notion of weak consistency; recognizing the weakness leads naturally to the notion of strong consistency.

Let $(X_{obj}S_{obj}, p_{obj})$ be a probability space and assume that S_{obj} is a finite Boolean algebra having $n < \infty$ elements. Let X_n be any set having n elements and $f\colon S_{obj} \to X_n$ be a bijection. Let p be any probability measure on the power set $\mathcal{P}(X_n)$ of X_n such that $p(A) \neq 0$ for any $\emptyset \neq A \in \mathcal{P}(X_n)$. Consider the standard product probability space

$$(X_{obj} \times X_n, S_{obj} \otimes \mathcal{P}(X_n), p_{obj} \times p) \tag{1.9}$$

of the probability spaces $(X_{obj}, S_{obj}, p_{obj})$ and $(X_n, \mathcal{P}(X_n), p)$, where $p_{obj} \times p$ is the product measure on $S_{obj} \otimes \mathcal{P}(X_n)$. Recall that the Boolean algebra $S_{obj} \otimes \mathcal{P}(X_n)$ is the smallest Boolean algebra on $X_{obj} \times X_n$ that contains all the sets of the form $A \times B$ with $A \in S_{obj}, B \in \mathcal{P}(X_n)$; and for $A \times B$, with $A \in S_{obj}$ and $B \in \mathcal{P}(X_n)$ we have

$$(p_{obj} \times p)(A \times B) = p_{obj}(A)p(B) \tag{1.10}$$

The maps h and g defined by

$$S_{obj} \ni A \mapsto h(A) \doteq (A \times X_n) \in S_{obj} \otimes \mathcal{P}(X_n) \tag{1.11}$$

$$\mathcal{P}(X_n) \ni B \mapsto g(B) \doteq (X_{obj} \times B) \in S_{obj} \otimes \mathcal{P}(X_n) \tag{1.12}$$

are Boolean algebra embeddings and $p_{subj} \doteq (p_{obj} \times p)$ is a probability function that satisfies Eq. (1.8) with

$$A' = X_{obj} \times \{f(A)\} \tag{1.13}$$

because

$$p_{subj}(h(A)|A') = \frac{p_{subj}((A \times X_n) \cap (X_{obj} \times \{f(A)\}))}{p_{subj}(X_{obj} \times \{f(A)\})} = \frac{(p_{obj} \times p)(A \times \{f(A)\})}{(p_{obj} \times p)(X_{obj} \times \{f(A)\})} \tag{1.14}$$

$$= \frac{p_{obj}(A)p(\{f(A)\})}{p_{obj}(X_{obj})p(\{f(A)\})} = p_{obj}(A) \tag{1.15}$$

In short, given $(X_{obj}, S_{obj}, p_{obj})$, the product probability space (1.9) satisfies the conditions required of $(X_{subj}, S_{subj}, p_{subj})$ by the definition of weak consistency of the Abstract Principal Principle, and the reason for this is that the restriction of the probability measure $p_{subj} = (p_{obj} \times p)$ to the (isomorphic copy $h(S_{obj})$ of) Boolean algebra S_{obj} of random events *coincides* with the probability measure p_{obj} giving the objective probability of the random events, and, as the calculation (1.14) shows, conditionalizing with respect to the propositions $A' = X_{obj} \times \{f(A)\}$ (which are interpreted as stating the values of objective probabilities) does not change $p_{subj}(A)$ because the propositions A' and $A \leftrightarrow h(A) = (A \times X_n)$, in virtue of their lying in different components of the product Boolean algebra, are independent with respect to the product measure $p_{subj} = (p_{obj} \times p)$ on the product algebra. That is to say, in the situation represented by this product state case, the agent's degrees of belief $p_{subj}(A)$ are equal to the objective probabilities $p_{obj}(A)$ *without* any conditionalizing, and conditionalizing them on the independent propositions stating the values of the objective probabilities does not change the already correct degrees of belief. Clearly, this is a very exceptional situation however, and it is more realistic to assume that the agent's degrees of belief are only equal to the objective probabilities *after* conditionalizing them on knowing the values of objective probabilities but *not* before. One would like to know if the Abstract Principal Principle is possible to maintain (is consistent) under these more realistic circumstances. The definition of weak consistency does not say anything about this more stringent consistency however; thus it is desirable to strengthen it in the manner specified in the next section.

1.4 Strong Consistency of the Abstract Principal Principle

Definition 4.1 The Abstract Principal Principle is defined to be *strongly consistent* if the following hold: Given any probability space $(X_{obj}, S_{obj}, p_{obj})$ and another probability measure p^0_{subj} on S_{obj}, there exists a probability space $(X_{subj}, S_{subj}, p_{subj})$ and a Boolean algebra embedding h of S_{obj} into S_{subj} such that

(i) For every $A \in S_{obj}$ there exists an $A' \in S_{subj}$ with the property

$$p_{subj}(h(A)|A') = p_{obj}(A) \tag{1.16}$$

(ii) If $A, B \in S_{obj}$ and $A \neq B$ then $A' \neq B'$.

(iii) The probability space $(X_{subj}, S_{subj}, p_{subj})$ is an extension of the probability space $(X_{obj}, S_{obj}, p^0_{subj})$ with respect to h; i.e. we have

$$p_{subj}(h(A)) = p^0_{subj}(A) \qquad A \in S_{obj} \tag{1.17}$$

The above Definition 4.1 differs from the definition of weak consistency (Definition 3.1) only by the additional requirement (iii). The intuitive content of this requirement is that the agent's prior probability function p_{subj} restricted to the random events can be equal to an arbitrarily chosen measure p^0_{subj} on S_{obj}; in particular the agent's prior subjective probabilities about random events can differ from the objective probabilities of the random events given by p_{obj}.

Definition 4.2 An agent's prior degrees of beliefs represented by p_{subj}^0 are called *non-extreme* (with respect to the objective probability function p_{obj}) if they satisfy the two conditions below:

- The agent's prior probabilities are *not* zero in events that have *non-zero* objective probability:

$$p_{subj}^0(A) = 0 \text{ entails } p_{obj}(A) = 0$$

- The agent's prior probabilities are *not* equal to 1 in events whose non-occurrence have a *nonzero* objective probability:

$$p_{subj}^0(A) = 1 \text{ entails } p_{obj}(A) = 1$$

Remark 4.3 An agent's prior degrees of beliefs represented by p_{subj}^0 are non-extreme if and only if the probability measure p_{obj} is *absolutely continuous* with respect to p_{subj}^0.

Proposition 4.4 *The Abstract Principal Principle is strongly consistent under the further assumption that the agent's prior degrees of beliefs are not extreme.*

We prove the above proposition in the Appendix (see Proposition A.1). Proposition 4.4 has already been proved in Gyenis and Rédei (2016) for finite probability spaces using a technique that constructs, in finite steps, an extension of a probability space with a finite Boolean algebra; the extension itself is a probability space in which the Boolean algebra has a finite number of elements. Proposition 4.4 claims more: the probability space can be infinite, as well. We apply a method which has been introduced in Gyenis and Rédei (2011).

The extension procedure differs from taking the standard product of the probability measure space to be extended with a suitable chosen other one, which was the technique used to prove the weak consistency of the Abstract Principal Principle (Proposition 3.2), and which also is the technique we will use to prove the weak consistency of the Stable Abstract Principal Principle later. We conjecture that strong consistency cannot be proved by taking the standard product as an extension.

1.5 Strengthening the Abstract Principal Principle: The Stable Abstract Principal Principle

Once the agent has learned the value of an objective probability and has adjusted his subjective degree of belief by conditionalizing on this evidence, $p_{subj}(h(A)|\ulcorner p_{obj}(A) = r\urcorner) = r$, he may then learn the value of another objective probability, $\ulcorner p_{obj}(B) = s\urcorner$, in which case he must conditionalize again. What should be the result of this second conditionalization? Since the agent's conditional degrees of belief $p_{subj}(h(A)|\ulcorner p_{obj}(A) = r\urcorner)$ in A are already correct, i.e. equal to the objective probabilities, it would be an irrational move on the agents's part to change his already correct degree of belief about A upon learning an additional *truth*, namely the value of the objective probability $p_{obj}(B)$. That is to say, a *rational* agent's conditional subjective degrees of belief should be *stable* in the sense of satisfying the following condition:

$$p_{subj}\left(h(A)|\ulcorner p_{obj}(A) = r\urcorner\right) = p_{subj}\left(h(A)|\ulcorner p_{obj}(A) = r\urcorner \cap \ulcorner p_{obj}(B) = s\urcorner\right) \quad (\forall B \in \mathcal{S}_{obj})$$
(1.18)

Another reason why stability should be a feature of the conditional subjective degrees of belief is the following. If A and B are independent with respect to their objective probabilities $p_{obj}(A \cap B) = p_{obj}(A)p_{obj}(B)$, then if the conditional subjective degrees of belief are stable in the sense of (1.18), then (assuming the Abstract Principal Principle) it follows immediately (see Gyenis and Rédei (2016) for more details) that if A and B are objectively independent, then they (their isomorphic images $h(A), h(B)$) are also subjectively independent: they are independent also with respect to the probability measure that represents *conditional* subjective degrees of belief, where the condition is that the agent knows the objective probabilities of *all* of A, B and $(A \cap B)$. In other words, in this case the conditional subjective degrees of beliefs properly reflect the objective independence relations of random events – they are *independence-faithful*. To put this negatively: if a subjective probability measure satisfying the Abstract Principal Principle is *not* stable, then, although the agent's degrees of belief are equal to the objective probabilities of individual random events after a single conditionalization on the values of the objective probabilities of these individual events, these (unstable) individually conditionalized subjective probability measures do *not* necessarily reflect the objective independence relations between the random events – stable conditionalized subjective probabilities reflect the objective probabilities more faithfully than unstable ones. Note that for the subjective degrees of belief to satisfy the independence-faithfulness condition, it is sufficient that stability (1.18) only holds for the restricted set of elements B in the Boolean subalgebra $\mathcal{S}_{obj}^{A,ind}$ of $\mathcal{S}_{obj}$ generated by the elements in $\mathcal{S}_{obj}$ that are independent of A with respect to p_{obj}.

All of this motivates an amendment to the Abstract Principal Principle to require stability of the subjective probabilities, resulting in the "Stable Abstract Principal Principle":

Stable Abstract Principal Principle The subjective probabilities $p_{subj}(A)$ are related to the objective probabilities $p_{obj}(A)$ as

$$p_{subj}(h(A)|\ulcorner p_{obj}(A) = r\urcorner) = p_{obj}(A)$$
(1.19)

Furthermore, the subjective probability function is *stable* in the sense that the following holds:

$$p_{subj}\left(h(A)|\ulcorner p_{obj}(A) = r\urcorner\right) = p_{subj}\left(h(A)|\ulcorner p_{obj}(A) = r\urcorner \cap \ulcorner p_{obj}(B) = s\urcorner\right) \quad (\forall B \in \mathcal{S}_{obj})$$
(1.20)

If the subjective probability function is only *independence-stable* in the sense that (1.20) above holds for all $B \in \mathcal{S}_{obj}^{A,ind}$, then the corresponding Stable Abstract Principal Principle is called the *Independence-Stable* Abstract Principal Principle.

The next section raises the problem of the consistency of the Stable Abstract Principal Principle.

1.6 Is the Stable Abstract Principal Principle Strongly Consistent?

Definition 6.1 The Stable Abstract Principal Principle is defined to be *weakly consistent* if it is weakly consistent in the sense of Definition 3.1 and the subjective probability function p_{subj} is stable: it satisfies condition (1.20). The *Independence-Stable* Abstract Principal Principle is defined to be weakly consistent if it is weakly consistent in the sense of Definition 3.1 and the subjective probability function p_{subj} is *independence-stable*: it satisfies (1.20) for all $B \in \mathcal{S}_{obj}^{A,ind}$.

Thus the problem of weak consistency of the Stable Abstract Principal Principle emerges, and we have

Proposition 6.2 *The Stable Abstract Principal Principle is weakly consistent.*

The above proposition entails in particular

Proposition 6.3 *The Independence-Stable Abstract Principal Principle is weakly consistent.*

We prove Proposition 6.2 in the Appendix. The proof is based on the product extension technique that was used to show the weak consistency of the Abstract Principal Principle in the category of probability spaces with a finite Boolean algebra, and so the proof reveals the same weakness of the notion of weak consistency of the Stable Abstract Principal Principle that we pointed out earlier in connection with the Abstract Principal Principle: In the situation represented by the product extension case, the agent's degrees of beliefs $p_{subj}(A)$ are equal to the objective probabilities $p_{obj}(A)$ *without* any conditionalizing, and conditionalizing them on the independent propositions stating the values of the objective probabilities does not change the already correct degrees of belief. Clearly, this is a very exceptional situation however, and it is more realistic to assume that the agent's degrees of belief are equal to the objective probabilities only *after* conditionalizing them on knowing the values of objective probabilities and not *before*. One would like to know if the (Independence-)Stable Abstract Principal Principle is consistent under these more realistic circumstances. The definition of weak consistency of the (Independence-)Stable Abstract Principal Principle does not say anything about this more stringent consistency however; thus it is desirable to strengthen it in the manner specified in the next definition.

Definition 6.4 The Stable Abstract Principal Principle is defined to be *strongly consistent* if it is strongly consistent in the sense of Definition 4.1 and the subjective probability function p_{subj} is stable. Explicitly:

The Stable Abstract Principal Principle is strongly consistent if the following hold: Given any probability space $(X_{obj}, S_{obj}, p_{obj})$ and another probability measure p^0_{subj} on S_{obj}, there exists a probability space $(X_{subj}, S_{subj}, p_{subj})$ and a Boolean algebra embedding h of S_{obj} into S_{subj} such that

(i) For every $A \in S_{obj}$ there exists an $A' \in S_{subj}$ with the property

$$p_{subj}(h(A)|A') = p_{obj}(A) \tag{1.21}$$

(ii) If $A, B \in S_{obj}$ and $A \neq B$ then $A' \neq B'$.

(iii) The probability space $(X_{subj}, S_{subj}, p_{subj})$ is an extension of the probability space $(X_{obj}, S_{obj}, p^0_{subj})$ with respect to h; i.e. we have

$$p_{subj}(h(A)) = p^0_{subj}(A) \qquad A \in S_{obj} \tag{1.22}$$

(iv) For all $B \in S_{obj}$ we have

$$p_{subj}(h(A)|A') = p_{subj}(h(A)|A' \cap B') \tag{1.23}$$

The *Independence-Stable* Abstract Principal principle is strongly consistent if (i)–(iii) above holds, and instead of (iv), we have

(iv')

$$p_{subj}(h(A)|A') = p_{subj}(h(A)|A' \cap B') \qquad \forall B \in S^{A,ind}_{obj} \tag{1.24}$$

Problem 6.5 Is the (Independence-)Stable Abstract Principal Principle strongly consistent?

The problem of strong consistency of both the Stable and the Independence-Stable Abstract Principal Principle is proved by Bana in Bana (2016) using techniques that are different from the one in Gyenis and Rédei (2016) that prove the weaker consistencies. That Bana's proof needs different techniques and that the method used in Gyenis and Rédei (2016) proving the strong consistency of the Abstract Principal Principle does *not* prove the strong consistency of the *Stable* Abstract Principal Principle can be shown by an explicit counterexample detailed in the Appendix (Proposition A.3). Specifically, one can show by an explicit calculation that if a Bayesian agent is in the very simple epistemic situation of having to form degrees of belief about the two events occurring in coin flipping, which is described by the simplest non-trivial Boolean algebra S^4_{obj} formed by the four events $\{\emptyset, A(Head), A^{\perp}(Tail), I\}$ with objective probabilities $p_{obj}(A) = \frac{1}{5}$ and $p_{obj}(A^{\perp}) = \frac{4}{5}$, then, if a Bayesian agent's prior subjective degrees of beliefs are $p^0_{subj}(A) = \frac{1}{3}$ and $p^0_{subj}(A^{\perp}) = \frac{2}{3}$, then the subjective degree of belief p_{subj} on the Boolean algebra

S_{subj} constructed using the method in the proof of Proposition 4.4 will *not* be stable. Note that we do *not* claim, that in such an epistemic situation it is not possible for the Bayesian agent to form a S_{subj} and extend p^0_{subj} to S_{subj} in such a way that the Abstract Principal Principle holds for p_{subj} possibly together with stability. We only claim that it is not possible to do this in the particular way in Gyenis and Rédei (2016) that proves in general the strong consistency of the Abstract Principal Principle. This indicates that it matters a lot how the Bayesian Agent extends the probability measure space containing the objective random events in order to follow the Bayesian norm specified by the Abstract principal Principle.

In the next two sections we argue that proving that the (Independence-)Stable Principal Principle is strongly consistent has ramifications both for the original Principal Principle and for Bayesianism in general.

1.7   Can Bayesian Agents Always Be Rational?

Call a Bayesian agent *omniscient* if he never has to adjust his subjective degrees of belief because his prior degrees of belief about random events in S_{obj} coincide with the objective probabilities p_{obj} on S_{obj} in all epistemic situations without any conditionalization: no matter what the set of random events S_{obj} and their objective probabilities p_{obj} are, facing (S_{obj}, p_{obj}), the agent has prior degrees of belief p^0_{subj} on S_{obj} such that

$$p^0_{subj}(A) = p_{obj}(A) \qquad A \in S_{obj} \tag{1.25}$$

The weak consistency of the Stable Abstract Principal Principle already shows that such an omniscient agent can always be a *rational* Bayesian agent in the sense that (i) he can always form a logically well-behaving set of propositions which is closed with respect to Boolean operations and which contains *both* the propositions stating that events in S happen *and* the propositions about the values of their objective probability; furthermore (ii) he can always have consistent degrees of belief as probabilities about *all* these propositions in such a way that (iii) if he conditionalizes his already correct prior degrees of belief on the values of any of the objective probabilities, his correct degrees of belief do not change. In short: weak consistency of the Stable Abstract Principal Principle ensures that an *omniscient* Bayesian agent can in principle be *always* rational.

We regard the strong consistency of the Stable Abstract Principal Principle necessary for a *non-omniscient* (hence more realistic) Bayesian agent to be able to have *rational* degrees of belief *in all epistemic situations*. For if the strong consistency of the Stable Abstract Principal Principle did *not* hold, then there would exist in principle epistemic situations the Bayesian agent could find himself in, in which at least one of the following (i)–(iii) *cannot* be maintained:

(i) Degrees of beliefs of the agent are represented by probability measures satisfying the usual axioms of measure theoretic probability.

(ii) The agent's prior degrees of beliefs differ from the objective probabilities but by learning the correct objective probabilities the agent can adjust his degrees of beliefs using Bayesian conditionalization so that they become equal to the objective probabilities.

(iii) The adjusted degrees of beliefs are stable: by learning additional truths about objective probabilities and re-conditionalizing his correct degrees of belief on them, the agent is not losing his already correct degrees of belief.

If not even the *Independence-Stable* Abstract Principal Principle could be proved to be strongly consistent, then (iii) above can be replaced with

(iii') The adjusted degrees of beliefs are stable: by learning additional truths about objective probabilities of *objectively independent events* and re-conditionalizing his correct degrees of belief on them, the agent is not losing his already correct degrees of belief (which also entails that the agent's conditioned degrees of belief reflect the objective independence properties of random events).

Thus the strong consistency of both the Stable and the Independence-Stable Abstract Principal Principle entails that a non-omniscient Bayesian agents can *always* be rational in principle.

1.8   Relation to Other Works

The idea of the Abstract Principal Principle can be traced back to Reichenbach's "Straight Rule" of induction connecting subjective and objective probabilities (the paper Eberhardt and Glymour (2009) gives a comprehensive review of Reichenbach's inductive logic and the role of the straight rule in it). The first claim about a possible inconsistency of the Straight Rule seems to be Miller's Paradox (Miller 1966); since Miller's work, the Straight Rule in the form of Eq. (1.1) is also called "Miller's Principle" and "Minimal Principle" (Vranas 2004; Pettigrew 2012). The inconsistency claim by Miller did not have anything to do with the type of consistency investigated in the present paper and Miller's Paradox was shown to be a pseudo-paradox resulting from the ambiguity of the formulation of the Straight Rule (Mackie 1966; Bub and Radner 1968; Howson and Oddie 1979).

1.8.1   David Lewis' Principal Principle and Strong Consistency of the Stable Abstract Principal Principle

Lewis (1986) introduced the term "Principal Principle" to refer to the specific principle that links subjective beliefs to chances in the manner expressed by (1.1): In the context of the Principal Principle $p_{subj}(A)$ is called the "credence", $Cr_t(A)$, of the agent in event A at time t, $p_{obj}(A)$ is the chance $Ch_t(A)$ of the event A at time t, and the Principal Principle is the stipulation that credences and chances are related as

$$Cr_t(A|\ulcorner Ch_t(A) = r\urcorner \cap E) = Ch_t(A) = r \qquad (1.26)$$

where E is any *admissible* evidence the agent has at time t in addition to knowing the value of the chance of A.

The relevance of the strong consistency of the (Independence-)Stable Abstract Principal Principle for Lewis' Principal Principle should now be clear: Clearly, the proposition $\ulcorner Ch_t(A) = r\urcorner$ is admissible evidence for (1.26), and substituting $E = \ulcorner Ch_t(A) = r\urcorner$ into Eq. (1.26) we obtain

$$Cr_t(A|\ulcorner Ch_t(A) = r\urcorner) = Ch_t(A) = r \qquad (1.27)$$

which, at any given time t, is an instance of the Abstract Principal Principle if we make the identifications $p_{obj}(A) = Ch_t(A)$, $p_{subj}(A) = Cr_t(A)$. We know from the proposition stating the strong consistency of the Abstract Principal Principle (Proposition 4.4) that, for any time parameter t, relation (1.27) can consistently hold in the sense that the agent can have any (non-extreme) prior credence about the objective chances of the events A, if at time t he is told the real values $Ch_t(A) = r$ of the chances (proposition $\ulcorner Ch_t(A) = r\urcorner$), then he can form a Boolean algebra containing both this information and the Boolean algebra of objective random events, he can extend his prior credence to the extended algebra to obtain a probability measure representing his credence, and he can perform a conditionalization of his credence in the enlarged probability space to make his conditioned credence equal to the objective chance. If, however, admissibility of evidence E is defined in such a way that propositions stating the values of chances of other events B at time t (i.e. propositions of the form $\ulcorner Ch_t(B) = s\urcorner$) are admitted as E, then (1.26) together with (1.27) entail that we also should have

$$Cr_t(A|\ulcorner Ch_t(A) = r\urcorner \cap \ulcorner Ch_t(B) = s\urcorner) = Ch_t(A) = r \qquad (1.28)$$

The relation (1.28) together with Eq. (1.27) is, at any given time t, an instance of the *Stable* Abstract Principal Principle if we make the identifications $p_{obj}(A) = Ch_t(A)$, $p_{subj}(A) = Cr_t(A)$ and $p_{obj}(B) = Ch_t(B)$. Thus the question of whether relations (1.28) and (1.27) can hold at all no matter what the prior credence of the agent and irrespective of what the objective chances are is exactly the question of whether the *Stable* Abstract Principal Principle is strongly consistent. If one allows as evidence E in (1.28) only propositions stating the value of objective chances of events B that are *objectively independent* of A, then the question of whether relations (1.28) and (1.27) can hold no matter what the prior credence of the agent and irrespective of what the objective chances are is exactly the question of whether the *Independence*-Stable Abstract Principal Principle is strongly consistent. Thus the consistency of Lewis' Principal Principle becomes an open question without proving the strong consistency of the Stable Abstract Principal Principle as soon as as one admits as evidence propositions stating the values of objective chances (probabilities) of random events. Since Lewis regarded admissible all propositions containing information that is "irrelevant" for the chance of A (Lewis 1986, p. 91), for Lewis, admissible evidence should include propositions about values of chances of events that are independent of A with respect to the probability measure describing their chances. Under this interpretation of "irrelevant" information, proving consistency the the *Independence*-Stable Abstract Principal Principle is necessary for Lewis' Principal principle to be consistent with classical probability theory. It should be emphasized that this kind of consistency has nothing to do with any metaphysics about chances or with the concept of natural law that one may have in the background of the Principal Principle; in particular, this inconsistency is different from the one related to "undermining" (see Sect. 1.8.2 below). This consistency expresses a simple but fundamental compatibility of the Principal Principle with the basic structure of probability theory.

1.8.2   Undermining and Consistency of Debugged Principal Principles

Lewis himself saw a consistency problem in connection with the Principal Principle (he called it the "Big Bad Bug"): If A is an event in the future of t that has a non-zero chance $r > 0$ of happening at that later time but we have knowledge E about the future that entails that A will in fact not happen, $E \subset A^{\perp}$, then substituting this E into (1.26) leads to contradiction if $r > 0$. Such an A is called an "unactualized future that undermines present chances" – hence the phrase "undermining" to refer to this situation. Since certain metaphysically motivated arguments based on a Humean understanding of chance led Lewis to think that one is forced to admit such an evidence E, he tried to "debug" the Principal Principle (Lewis 1994); the same sort of debugging was proposed simultaneously by Hall (1994) and Thau (1994). A number of other debugging attempts, modifications and analysis have followed Black (1998), Roberts (2001), Loewer (2004), Hall (2004), Hoefer (2007), Ismael (2008), Meacham (2010), Glynn (2010), Nissan-Rozen (2013), Pettigrew (2013), Frigg and Hoefer (2015), and to date no consensus has emerged as to which of the debugged versions of Lewis original Principal Principle is tenable: Vranas claims (2004) that there was no need for a debugging in the first place; Briggs (2009) argues that none of the modified principles work; Pettigrew (2012) provides a systematic framework that allows in principle to choose the correct Principal Principle depending on one's metaphysical concept of chance.

Papers aiming at "debugging" Lewis' Principal Principle typically combine the following three moves (a), (b) or (c):

(a) Restricting the admissible evidence in (1.26) to a particular class $\mathcal{A}_A$ of propositions in order to avoid inconsistency arising from "undermining" (Hoefer 2007).

(b) Modifying the Principal Principle by replacing the objective chance $Ch_t(A)$ of A on the right hand side of (1.26) with a value $F(A)$ given by a function F different from the objective chance (probability) function (New Principle by Hall 1994; General Principal Principle by Lewis 1980 and by Roberts 2001).

(c) Modifying the Principal Principle by replacing the conditioning proposition $\ulcorner Ch_t(A) = r \urcorner \cap E$ on the left hand side of (1.26) by a different conditioning proposition C_A, which is a conjunction of some propositions from S_{obj}, $\mathcal{A}_A$, and propositions of form $\ulcorner p_{obj}(B) = r \urcorner$ (Conditional Principle and General Principle by Vranas 2004); General Recipe by Ismael (2008).

To establish a theory of chance along a debugging strategy characterized by a combination of (a), (b) and (c), it is not enough to show however that undermining is avoided: one has to prove that the debugged Principal Principle is consistent in the sense of Definition 8.1 below, which is in the spirit of the notion consistency investigated in this paper:

Definition 8.1 We say that the "$(\mathcal{A}_A, C_A, F)$-debugged" Principal Principle is *strongly* consistent if the following hold:

Given any probability space $(X_{obj}, S_{obj}, p_{obj})$ and another probability measure p^0_{subj} on S_{obj}, there exists a probability space $(X_{subj}, S_{subj}, p_{subj})$ and a Boolean algebra embedding h of S_{obj} into S_{subj} such that

(i) For every $A \in S_{obj}$ the set $\mathcal{A}_A$ is in S_{subj}, and for every $A \in S_{obj}$ there exists a $C_A \in S_{subj}$ with the property

$$p_{subj}(h(A)|C_A) = F(A) \qquad (1.29)$$

(ii) If $A, B \in S_{obj}$ and $A \neq B$ then $C_A \neq C_B$.

(iii) The probability space $(X_{subj}, S_{subj}, p_{subj})$ is an extension of the probability space $(X_{obj}, S_{obj}, p^0_{subj})$ with respect to h; i.e. we have

$$p_{subj}(h(A)) = p^0_{subj}(A) \qquad A \in S_{obj} \qquad (1.30)$$

(iv) For all $A \in S_{obj}$ and for all $B \in \mathcal{A}_A$ we have

$$p_{subj}(h(A)|C_A) = p_{subj}(h(A)|C_A \cap B) \qquad (1.31)$$

We say that the "$(\mathcal{A}_A, C_A, F)$-debugged" Principal Principle is *weakly* consistent if (i), (ii) and (iv) hold.

The above definition of weak and strong consistency of the $(\mathcal{A}_A, C_A, F)$-debugged Principal Principle is completely analogous to the definition of weak and strong consistency of the Stable Abstract Principal Principle (Definitions 6.1 and 6.4). The differences are that (i) the right hand sides of (1.8) and (1.21) are replaced by $F(A)$ in (1.29); (ii) the conditioning events in (1.29) are allowed to be more complex propositions than $\ulcorner p_{obj}(A) = r \urcorner$ in (1.8) and (1.21); and (iii) stability (1.23) is required in (1.31) for propositions B representing admissible evidence, which can now be propositions different from propositions stating the values of chances of events different from A. Taking F as the objective probability p_{obj}, C_A as $\ulcorner p_{obj}(A) = r \urcorner$, and allowing $\mathcal{A}_A$ to be the set of propositions stating probabilities of events $B \neq A$ one recovers Definitions 6.1 and 6.4 as particular cases of Definition 8.1. On the other hand, taking specific C_A, and F, one obtains particular consistency definitions expressing the consistency of specific debugged Principal Principles. For instance, stipulations

$$C_A = B \cap \ulcorner p_{obj}(A|B) = r \urcorner \qquad (1.32)$$

$$F(A) = p_{obj}(A) \qquad (1.33)$$

yield Vranas' Conditional Principle (Vranas 2004, p. 370); whereas Hall's New Principle (Hall 1994, p. 511) can be obtained by

$$C_A = H_{t,w} \cap T_w \qquad (1.34)$$

$$F(A) = p_{obj}(A|T_w) \qquad (1.35)$$

where $H_{t,w}$ is "the proposition that completely characterizes w's history up to time t" (Hall 1994, p. 506) and T_w is the "proposition that completely characterizes the laws at w" (Hall 1994, p. 506) (w being a possible world).

Are the $(\mathcal{A}_A, C_A, F)$-debugged Principal Principles weakly or strongly consistent for the F's, C_A's and $\mathcal{A}_A$ proposed in the literature? Proving this type of consistencies is necessary for the respective debugged Principal Principles to be compatible with measure theoretic probability theory, including Bayesian conditionalization. To our best knowledge such consistency proofs have *not* been given: it seems that this type of consistency is tacitly assumed in the works analyzing the modified Principal Principles, although, as the propositions and their proofs presented in this paper show, the truth of these types of consistency claims are far from obvious.

1.8.3   Gaifman's Theory of Higher Order Probabilities and Strong Consistency of the Stable Abstract Principal Principle

The problem of strong consistency of the Stable Abstract Principle is relevant from the perspective of existence of particular models of the axioms of higher order probability theory (HOP) suggested by Gaifman (1988). To explain the relevance, we recall here Gaifman's axiomatic specification of a simple HOP.

Let $\mathcal{I}$ be a set of closed sub-intervals of $[0, 1]$ containing $[0, 1]$ and the empty set.

Definition 8.2 The quadruple $(X, \mathcal{S}, p, pr)$ is called a higher order probability space if the following hold:

1. $(X, \mathcal{S}, p)$ is a probability space
2. pr is a map

$$pr: \mathcal{S} \times \mathcal{I} \to \mathcal{S} \tag{1.36}$$

having properties (I)–(VI) below:

(I) $pr(A, \emptyset) = \emptyset$ for all $A \in \mathcal{S}$
(II) $pr(A, [0, 1]) = X$ for all $A \in \mathcal{S}$
(III) if d_1, d_2 and $d_1 \cup d_2$ are all intervals, then for all $A \in \mathcal{S}_0$

$$pr(A, d_1 \cup d_2) = pr(A, d_1) \cup pr(A, d_2) \tag{1.37}$$

(IV) If $d_i \ (i = 1, 2, \ldots n \ldots)$ is a countable set of intervals from $\mathcal{I}$, then

$$\cap_i pr(A, d_i) = pr(A, \cap_i d_i) \tag{1.38}$$

(V) If $A_i \ (i = 1, 2, \ldots n \ldots)$ is a countable set of pairwise disjoint elements from $\mathcal{S}$, and $[\alpha_i, \beta_i]$ is a countable set of intervals from $\mathcal{I}$, and $[\alpha, \beta]$ is the interval with $\alpha = \sup_i \alpha_i$ and $\beta = \sup_i \beta_i$, then

$$\cup_i pr(A_i, [\alpha_i, \beta_i]) \subseteq pr(\cup A_i, [\alpha, \beta]) \tag{1.39}$$

(VI) If $C \in \mathcal{S}$ is a finite intersection of the form

$$C = pr(A_1, d_1) \cap pr(A_2, d_2) \cap \ldots \cap pr(A_n, d_n) \tag{1.40}$$

then for all $A \in \mathcal{S}$ for which $p(pr(A, d) \cap C) \neq 0$ we have

$$p(A|pr(A, d) \cap C) \in d \tag{1.41}$$

The interpretation of a HOP is the following: For any event $A \in \mathcal{S}$ and any interval $d \in \mathcal{I}$, $pr(A, d)$ is the event that the (objective, "true") probability of A lies in the interval d. Features (I)–(V) essentially stipulate that this interpretation of pr is in harmony with measure theoretic probability; whereas axiom (VI) connects pr and the "subjective" probability p in the spirit of the Stable Abstract Principal Principle: Eq. (1.41) says that the conditional subjective probability of A on condition that its objective probability lies in the interval d, lies in the interval d – and that this conditional subjective probability is stable with respect to further conditionalization on propositions stating that other events A_i have objective probabilities lying in intervals d_i. If one restricts the closed intervals in $\mathcal{I}$ to one-point sets, (1.41) reduces to the form (1.20).

If one regards the theory of HOP as an axiomatic theory, then the question arises whether models of the theory exist. Gaifman provides a few specific examples that are models of the axioms (Gaifman 1988, p. 208–210) but he does not raise the general issue of what kind of models exist. What one would like to know is whether any objective probability theory can be made part of a HOP in such a way that the objective probabilities are related to the subjective ones in the manner of (1.41). To be more precise, the question is whether for any objective probability space $(X_{obj}, S_{obj}, p_{obj})$ with a prior subjective probability p^0_{subj} on S_{obj} there exists a HOP (X, S, p, pr) such that

- (X, S, p) is an extension of $(X_{obj}, S_{obj}, p^0_{subj})$
- For all $A \in S_{obj}$ and for $r = p_{obj}(A)$ we have

$$p(A|pr(A, [r, r]) \cap C) = r \qquad (1.42)$$

Clearly, proving the existence of such a HOP for all $(X_{obj}, S_{obj}, p_{obj})$ and for every prior subjective probability p^0_{subj} on S_{obj} would entail that the Stable Abstract Principal Principle is strongly consistent.

1.9 The General Principal Principle

In this section we extend the analysis of the Principal Principle to non-classical (quantum) probability spaces, and investigate in particular the consistency of what we call a General Principal Principle, which is formulated in terms of non-classical probability spaces. The extension is motivated by the fact that a Bayesian interpretation of quantum probabilities is a possible (Caves et al. 2002), although not uncontroversial (Rédei 1992; Valente 2007) position and that conditionalization in quantum probability theory and its relation to quantum Bayesianism also is an issue discussed in current literature (Leifer and Spekkens 2013).

Let $\mathcal{H}$ be a Hilbert space and denote by $\mathcal{B}(\mathcal{H})$ the C*-algebra of all bounded linear operators acting on $\mathcal{H}$. We assume that $\mathcal{H}$ is separable. If $\mathcal{M} \subseteq \mathcal{B}(\mathcal{H})$ is a von Neumann algebra, then the set of projections $\mathcal{P}(\mathcal{M})$ in $\mathcal{M}$ is a complete orthomodular lattice (Rédei 1998). A *generalized probability space* is a pair $(\mathcal{P}(\mathcal{M}), \phi)$, where $\mathcal{P}(\mathcal{M})$ is the set of projections of a von Neumann algebra $\mathcal{M}$ and ϕ is a normal state on $\mathcal{M}$ (a probability measure on $\mathcal{P}(\mathcal{M})$). For a short review of quantum probability theory see Rédei and Summers (2007).

If $\mathcal{M}$ and $\mathcal{N}$ are von Neumann algebras and $\mathcal{N}$ is a von Neumann subalgebra of $\mathcal{M}$ then a positive linear mapping $\mathbb{E} : \mathcal{M} \to \mathcal{N}$ is called a conditional expectation if it has the following two properties:

- $\mathbb{E}(I) = I$,
- $\mathbb{E}(R_1 S R_2) = R_1 \mathbb{E}(S) R_2$ whenever $R_1, R_2 \in \mathcal{N}$ and $S \in \mathcal{M}$.

In the context of generalized probability spaces the strong consistency of the Abstract Principal Principle (which we call then the General Principal Principle) can be defined analogously:

Definition 9.1 The General Principal Principle is *strongly consistent* if the following hold: Given any von Neumann algebra $\mathcal{N}_{obj}$ with two states ϕ_{obj} and ϕ_{subj} on $\mathcal{N}_{obj}$, (i)–(iii) and (A)–(B) below hold

(i) There exists a von Neumann algebra $\mathcal{N}_{obj+subj}$, a von Neumann subalgebra $\mathcal{N}_{subj}$ of $\mathcal{N}_{obj+subj}$ and an injective von Neumann algebra homomorphism $h : \mathcal{N}_{obj} \to \mathcal{N}_{obj+subj}$ such that

$$h(\mathcal{N}_{obj}) \cap \mathcal{N}_{subj} = \{0, I\} \qquad (1.43)$$

(ii) There exists a conditional expectation $\mathbb{E}$ from $\mathcal{N}_{obj+subj}$ to $\mathcal{N}_{subj}$.
(iii) There is a state $\phi_{obj+subj}$ on $\mathcal{N}_{obj+subj}$ with the properties (A) and (B) below.

    (A) State Extension: The restriction of $\phi_{obj+subj}$ to $\mathcal{N}_{obj}$ coincides with ϕ_{subj}.
    (B) Principle: The restriction to $\mathcal{N}_{obj}$ of the $\mathbb{E}$-conditioned state $(\phi_{obj+subj} \circ \mathbb{E})$ coincides with ϕ_{obj}.

The interpretation of Definition 9.1 is analogous to the interpretation of the definition of strong consistency of the Principal Principle in the context of classical probability theory:

- $\mathcal{N}_{obj}$ is the algebra of objective physical quantities with ϕ_{obj} being their objective expectation values. The state ϕ_{subj} yields the subjective expectation values.
- $\mathcal{N}_{obj+subj}$ is the algebra containing both the objective quantities $\mathcal{N}_{obj}$ (via h) and the algebra $\mathcal{N}_{subj}$ projections of which represent propositions that express subjective knowledge about the objective expectation values.
- $\mathbb{E}$ is the conditioning with respect to the knowledge/known values of expectations.
- $\phi_{obj+subj}$ is the probability measure describing subjective degrees of beliefs in objective matters *before* taking into account the real expectation values of quantities.
- The $\mathbb{E}$-conditioned state $(\phi_{obj+subj} \circ \mathbb{E})$ is the probability measure giving the subjective degrees of beliefs *after* taking into account, via conditionalization, the real expectation values.

Theorem 9.2 *The General Principal Principle is strongly consistent.*

We prove the above theorem in the Appendix (see Theorem A.4). The proof is based on the idea of the proof of strong consistency of the Abstract Principal Principle, cf. Proposition A.1.

One also can raise the problem of whether a *Stable* General Principal Principle also is maintainable. The motivation for (and significance of) such a Stable General Principal Principle are the same as in the context of classical probability theory. Since we do not have any result on the consistency of such a strengthened General

Principal Principle, we do not formulate it here explicitly. Investigations in this direction remains a task for the future.

Acknowledgements Research supported in part by the National Research, Development and Innovation Office, K 115593. We gratefully acknowledge support from the Bilateral Mobility Project "Probability, causality and determinism" of the Hungarian and Polish Academies of Sciences between 2012–2016.

Appendix

Proof of Strong Consistency of the Abstract Principal Principle (Proposition 4.4)

The statement of strong consistency of the Abstract Principal Principle follows from Proposition A.1 below if we make the following identifications:

- $(X_{obj}, S_{obj}, p_{obj}) \leftrightarrow (X, S, \hat{p})$
- $(X_{obj}, S_{obj}, p^0_{subj}) \leftrightarrow (X, S, p)$
- $(X_{subj}, S_{subj}, p_{subj}) \leftrightarrow (X', S', p')$

Proposition A.1 *Let (X, S, p) be a probability space and let $\hat{p}$ be another probability measure on S such that $\hat{p}$ is absolutely continuous with respect to p (cf. Definition 4.2). Then there exists an extension (X', S', p') of (X, S, p) with respect to the embedding $h: S \to S'$ having the following properties:*

(i) For all $A \in S$ there is $A' \in S'$ such that

$$p'(h(A)|A') = \hat{p}(A)$$

(ii) $A \neq B$ implies $A' \neq B'$

Proof We start by briefly recalling from Gyenis and Rédei (2011) the definition of ⊛-product of probability spaces, in a special case. For a detailed description we refer to the Appendix of Gyenis and Rédei (2011). Let us denote by $([0, 1], \mathcal{L}, \lambda)$ the standard Lebesgue measure space over the unit interval $[0, 1]$. The ⊛-product of the spaces (X, S, p) and $([0, 1], \mathcal{L}, \lambda)$ is a space

$$(X', S', p') = (X, S, p) \circledast ([0, 1], \mathcal{L}, \lambda),$$

which enjoys the following properties.

- Each element $f \in S'$ (can be identified with) a function $f : [0, 1] \to S$.
- Characteristic functions $\chi^A_B : [0, 1] \to S$, $(A \in S, B \in \mathcal{L})$ are elements of S'.

$$\chi_B^A(x) = \begin{cases} A & \text{if } x \in B \\ \emptyset & \text{otherwise.} \end{cases}$$

- Simple functions $f : [0, 1] \to S$ are all elements of S'. A function is simple, if there is a natural number n, a measurable partition $\bigsqcup_{i<n} B_i$ of $[0, 1]$, $B_i \in \mathcal{L}$ and events $A_i \in S$ such that $f(x) = A_i$ if and only if $x \in B_i$.
- The measure (probability) of a simple function f is defined as

$$p'(f) = \int_0^1 p(f(x)) \, d\lambda(x)$$

which, using the previous notation, is the same as

$$p'(f) = \sum_{i<n} p(A_i)\lambda(B_i).$$

- The mapping $h : S \to S', h(A) = \chi_{[0,1]}^A$ is an injective σ-algebra homomorphism.

So let (X', S', p') be the $\circledast$-product $(X, S, p) \circledast ([0, 1], \mathcal{L}, \lambda)$. The image $h(A)$ of an event $A \in S$ is then the constant function

$$h(A)(x) = A, \qquad x \in [0, 1].$$

Now, for each $A \in S$ we define $A' \in S'$. Let $\alpha \in [0, 1]$ be a real number the value of which will be chosen later and pick any Lebesgue-measurable set $B \subseteq [0, 1]$ with measure $\lambda(B) = \alpha$. Write A' for the function $A' : [0, 1] \to S$,

$$A'(x) = \begin{cases} A & \text{if } x \in B \\ A^\perp & \text{otherwise.} \end{cases}$$

Then A' is a simple function which belongs to S' and observe that

$$\big(h(A) \cap A'\big)(x) = \begin{cases} A & \text{if } x \in B \\ \emptyset & \text{otherwise.} \end{cases}$$

Using this observation we get

$$p'(h(A) \,|\, A') = \frac{p'(h(A) \cap A')}{p'(A')} = \frac{\int_0^1 p \circ \big(h(A) \cap A'\big) d\lambda}{\int_0^1 p \circ A' d\lambda}$$

$$= \frac{\displaystyle\int_B p(A)d\lambda + \int_{[0,1]\setminus B} p(\emptyset)d\lambda}{\displaystyle\int_B p(A)d\lambda + \int_{[0,1]\setminus B} p(A^\perp)d\lambda} = \frac{\lambda(B) \cdot p(A) + \lambda([0,1]\setminus B) \cdot p(\emptyset)}{\lambda(B) \cdot p(A) + \lambda([0,1]\setminus B) \cdot p(A^\perp)}$$

$$= \frac{\alpha \cdot p(A)}{\alpha \cdot p(A) + (1-\alpha) \cdot (1-p(A))}.$$

This means that if we choose α such that

$$\frac{\alpha \cdot p(A)}{\alpha \cdot p(A) + (1-\alpha) \cdot (1-p(A))} = \hat{p}(A), \tag{1.44}$$

then we get $p'(h(A) \mid A') = \hat{p}(A)$.

By our assumptions, if $p(A) = 1$ then $\hat{p}(A) = 1$ and thus any $\alpha \neq 0$ makes (1.44) true. Similarly, if $p(A) = 0$, then $\hat{p}(A) = 0$, which means that any $\alpha \neq 1$ will do. Also, if $\hat{p}(A) = 0$, then $\alpha = 0$ will do. Therefore we may assume $0 < p(A) < 1$ and $0 < \hat{p}(A) \leq 1$. By re-ordering Eq. (1.44) and using the notation $p = p(A), r = \hat{p}(A)$ we obtain the equation

$$\alpha = \frac{rp - r}{rp - r + pr - p} \tag{1.45}$$

In order to guarantee (1.44) we only have to show that α in Eq. (1.45) is between 0 and 1. To do so, observe that since $0 < p < 1$ and $0 < r \leq 1$ we have $rp < r$ and $pr \leq p$. This means that both the numerator and the denominator of the fraction in (1.45) is negative, whence α is positive. On the other hand, we have

$$0 \geq pr - p$$
$$rp - r \geq rp - r + pr - p$$
$$\frac{rp - r}{rp - r + pr - p} \leq 1$$

Thus we proved that $0 \leq \alpha \leq 1$ can always be chosen so that Eq. (1.44) holds. This completes the proof. □

Proof of Weak Consistency of the Stable Abstract Principal Principle (Proposition 6.2)

The statement of weak consistency of the Stable Abstract Principal Principle follows from Proposition A.2 below if we make the following identifications:

- $(X_{obj}, \mathcal{S}_{obj}, p_{obj}) \leftrightarrow (X, \mathcal{S}, p)$
- $(X_{subj}, \mathcal{S}_{subj}, p_{subj}) \leftrightarrow (X', \mathcal{S}', p')$

Proposition A.2 *Let $(X, \mathcal{S}, p)$ be a probability space. Then there exists an extension $(X', \mathcal{S}', p')$ of $(X, \mathcal{S}, p)$ with respect to a Boolean algebra homomorphism $h \colon \mathcal{S} \to \mathcal{S}'$ such that*

(i) For all $A \in \mathcal{S}$ there is $A' \in \mathcal{S}'$ such that

$$p'\big(h(A)|A'\big) = p(A)$$

(ii) $A \neq B$ implies $A' \neq B'$
(iii)

$$p'(h(A)|A') = p'\big(h(A)|A' \cap B'\big) \qquad (\forall B' \in \mathcal{S}) \tag{1.46}$$

Proof Let $(X, \mathcal{S}, p)$ be a probability space and Y_0 be a set disjoint from $\mathcal{S}$ and having the same cardinality as the cardinality of $\mathcal{S}$. We can think of Y_0 as having elements y_A labeled by elements $A \in \mathcal{S}$. Consider the set

$$Y \doteq Y_0 \cup \{y\} = \{y_A : A \in \mathcal{S}\} \cup \{y\}$$

where y is an auxiliary element different from every y_A. Take the powerset $\mathcal{P}(Y)$ and let q be any probability measure on $\mathcal{P}(Y)$ such that $q(\{y\}) \neq 0$. Then $(Y, \mathcal{P}(Y), q)$ is a probability space and we can form the product space

$$(X', \mathcal{S}', p') = (X \times Y, \mathcal{S} \otimes \mathcal{P}(Y), p \times q)$$

with $p' = (p \times q)$ being the product measure on $\mathcal{S} \otimes \mathcal{P}(Y)$. Recall that by definition of the product measure, for sets of the form $A \times B$ ($A \in \mathcal{S}$, $B \in \mathcal{P}(Y)$) we have

$$p'(A \times B) = p(A)q(B)$$

The map $h : \mathcal{S} \to \mathcal{S}'$ defined by $h(A) \doteq A \times Y$ is an injective, measure preserving Boolean algebra embedding. For each $A \in \mathcal{S}$ put

$$A' \doteq X \times \{y_A, y\}$$

It is clear that (ii) in the proposition holds for A', B' so defined.

To see (i) one can compute:

$$p'(h(A)|A') = \frac{p'(h(A) \cap A')}{p'(A')} = \frac{p'\big((A \times Y) \cap (X \times \{y_A, y\})\big)}{p'(X \times \{y_A, y\})}$$

$$= \frac{p'\big(A \times \{y_A, y\}\big)}{p'\big(X \times \{y_A, y\}\big)} = \frac{p(A) \cdot q(\{y_A, y\})}{p(X) \cdot q(\{y_A, y\})} = p(A)$$

Condition (iii) also holds because

$$
\begin{aligned}
p'(h(A)|A' \cap B') &= \frac{p'(h(A) \cap A' \cap B')}{p'(A' \cap B')} \\
&= \frac{p'\big((A \times Y) \cap (X \times \{y_A, y\}) \cap (X \times \{y_B, y\})\big)}{p'\big((X \times \{y_A, y\}) \cap (X \times \{y_B, y\})\big)} \\
&= \frac{p'(A \times \{y\})}{p'(X \times \{y\})} = \frac{p(A) \cdot q(\{y\})}{p(X) \cdot q(\{y\})} = p(A)
\end{aligned}
$$

$\square$

Example Showing That the General Method That Proves the Strong Consistency of the Abstract Principal Principle Does Not Prove the Strong Consistency of the Stable Abstract Principal Principle

Proposition A.3 *There exists a probability space* $(X, \mathcal{S}, \hat{p})$ *and a probability measure* p *on* $\mathcal{S}$ *such that the probability measure* p' *in the probability space* $(X', \mathcal{S}', p')$ *constructed in the proof of Proposition 4.4 will* not *be stable, i.e. for some* $A, B \in \mathcal{S}$ *we have:*

$$
p'(h(A)|A') \neq p'(h(A)|A' \cap B') \tag{1.47}
$$

Proof Let $\mathcal{S} = \mathcal{S}^4$ be the Boolean algebra having four elements $\{\emptyset, A, B, I\}$ (clearly: $B = A^{\perp}$) and let $\hat{p}$ be a probability measure on $\mathcal{S}^4$. Assume that $p(A) = x$ and $p(B) = p(A^{\perp}) = 1 - x$ for some real number $x \in [0, 1]$. Since $\mathcal{S}^4$ has only two non-trivial elements, constructing the space $(X', \mathcal{S}', p')$ that extends $(X, \mathcal{S}^4, p)$ in the way detailed in the proof of Proposition 5 in Gyenis and Rédei (2016) is finished after one has extended the probability space $(X, \mathcal{S}^4, \hat{p})$ twice in the manner described in the proof of Proposition 5 in Gyenis and Rédei (2016): First one constructs $(X_1, \mathcal{S}_1, p_1)$ with the Boolean algebra homomorphism $h_1: \mathcal{S}^4 \to \mathcal{S}_1$ in such a way that Eq. (1.44) hold with a suitable α_1. In the second step one constructs the extension $(X_2, \mathcal{S}_2, p_2)$ of $(X_1, \mathcal{S}_1, p_1)$ with the Boolean algebra homomorphism $h_2: \mathcal{S}_1 \to \mathcal{S}_2$ in such a way that (1.44) hold with $B = A^{\perp}$ in place of A_2 and with a suitable α_2. Following the notation in the proof of Proposition 5 in Gyenis and Rédei (2016), we can write:

$$
h = h_2 h_1 \tag{1.48}
$$

$$
A' = h_2(A^*) \tag{1.49}
$$

$$
B' = \big(h(B)\big)^* \tag{1.50}
$$

Then the equations expressing the Abstract Principal Principle in connection with events A and B can be written as (cf. Eq. (1.44))

$$p'\big(h(A)|A'\big) = \frac{\alpha_1 x}{\alpha_1 x + (1 - \alpha_1)(1 - x)} \tag{1.51}$$

$$p'\big(h(B)|B'\big) = \frac{\alpha_2(1 - x)}{\alpha_2(1 - x) + (1 - \alpha_2)x} \tag{1.52}$$

One can now compute explicitly the conditional probability $p'\big(h(A)|A' \cap B'\big)$ of $h(A)$ in the probability space $(X', \mathcal{S}', p')$ after a *second* conditionalization on the value of the objective probability probability of B. The computation yields

$$p'\big(h(A)|A' \cap B'\big) = \frac{(1 - \alpha_2)\alpha_1 x}{\alpha_2(1 - \alpha_1)(1 - x) + (1 - \alpha_2)\alpha_1 x} \tag{1.53}$$

Note that since $p(h(A)|A') = \hat{p}(A)$ and $p'(h(B)|B') = \hat{p}(B) = \hat{p}(A^{\perp})$, we have

$$p(h(A)|A') + p(h(B)|B') = 1 \tag{1.54}$$

hence if we take $x = \alpha_1 = \frac{1}{3}$, then we get

$$p(h(A)|A') = \frac{1}{5}$$

$$p(h(B)|B') = \frac{4}{5} \implies \alpha_2 = \frac{2}{3}$$

$$p(h(A)|A' \cap B') = \frac{1}{9}$$

Therefore $p(h(A)|A') \neq p(h(A)|A' \cap B')$.   □

Proof of Strong Consistency of the General Principal Principle (Proof of Theorem 9.2)

Theorem A.4 *The General Principal Principle is strongly consistent. In more details: Given any von Neumann algebra $\mathcal{N}_{obj}$ with two states ϕ_{obj} and ϕ_{subj} on $\mathcal{N}_{obj}$, (i)–(iii) and (A)–(B) below hold*

(i) *There exists a von Neumann algebra $\mathcal{N}_{obj+subj}$, a von Neumann subalgebra $\mathcal{N}_{subj}$ of $\mathcal{N}_{obj+subj}$ and an injective von Neumann algebra homomorphism $h : \mathcal{N}_{obj} \to \mathcal{N}_{obj+subj}$ such that*

$$h\big(\mathcal{N}_{obj}\big) \cap \mathcal{N}_{subj} = \{0, I\} \tag{1.55}$$

(ii) *There exists a conditional expectation $\mathbb{E}$ from $\mathcal{N}_{obj+subj}$ to $\mathcal{N}_{subj}$.*

(iii) *There is a state $\phi_{obj+subj}$ on $\mathcal{N}_{obj+subj}$ with the properties (A) and (B) below.*

(A) **State Extension:** *The restriction of $\phi_{obj+subj}$ to $\mathcal{N}_{obj}$ coincides with ϕ_{subj}.*

(B) **Principle:** *The restriction to $\mathcal{N}_{obj}$ of the $\mathbb{E}$-conditional state $(\phi_{obj+subj} \circ \mathbb{E})$ coincides with ϕ_{obj}.*

Proof Fix the von Neumann algebra $\mathcal{N}_{obj}$ acting on a separable Hilbert space $\mathcal{H}$, and take two normal states ϕ_{obj} and ϕ_{subj} on $\mathcal{N}_{obj} \subseteq \mathcal{B}(\mathcal{H})$.

Let us fix a classical probability space $\mathfrak{X} = (X, \Sigma, \mu)$, where X is a set, Σ is a σ-algebra of subsets of X and μ is a probability measure on Σ. Throughout we will have $\mathfrak{X} = ([0, 1], \Lambda, \lambda)$ in mind, where Λ is the σ-algebra of Lebesgue-measurable subsets of the unit interval, and λ is the Lebesgue measure. We start by recalling some definitions and facts from Hilbert space theory, e.g. from Fonseca and Leoni (2007), Kadison and Ringrose (1986), and Takesaki (2003).

Definition A.5 A function $s : X \to \mathcal{H}$ is called *simple* if it is of the following form:

$$s = \sum_{i=1}^{\ell} h_i \chi_{E_i},$$

where $\ell \in \mathbb{N}$, $h_i \in \mathcal{H}$ and the sets E_i form a partition of X such that each E_i is measurable ($E_i \in \Sigma$). χ_E is the characteristic function of the set E.

A function $u : X \to \mathcal{H}$ is called *strongly measurable*, if there is a sequence $\{s_n\}_{i \in \mathbb{N}}$ of simple functions such that

$$\lim \|s_n(x) - u(x)\|_{\mathcal{H}} = 0 \quad \text{for } \mu\text{-almost all } x \in X$$

If $u : X \to \mathcal{H}$ is strongly measurable, then the map $x \mapsto \|u(x)\|_{\mathcal{H}}$ is measurable in the classical sense, i.e. (Σ, Λ)-measurable.

Definition A.6 We define $\mathfrak{H} = L^2(\mathfrak{X}, \mathcal{H})$ as follows:

$$L^2(\mathfrak{X}, \mathcal{H}) = \left\{ u : X \to \mathcal{H} : u \text{ is strongly measurable and } \|u\|_{\mathfrak{H}} < \infty \right\},$$

where $\|u\|_{\mathfrak{H}}$ for a strongly measurable $u : X \to \mathcal{H}$ is defined as

$$\|u\|_{\mathfrak{H}} = \left(\int_X \|u(x)\|_{\mathcal{H}}^2 \, d\mu(x) \right)^{1/2}$$

As usual, we identify functions which are almost everywhere the same. If, for the elements $x, y \in \mathfrak{H}$ we let

$$(x, y) = \int_X (x(t), y(t))_{\mathcal{H}} \, d\mu(t),$$

then this defines a scalar product which generates a norm $\| \cdot \|_{\mathfrak{H}}$ in $\mathfrak{H}$.

By Theorem 2.110 of Fonseca and Leoni (2007), $(\mathfrak{H}, \|\cdot\|_{\mathfrak{H}})$ is a Banach space in which the set of simple functions is dense. Consequently, $\mathfrak{H}$ is a Hilbert space. Further, if X is a separable metric space, μ is a Radon measure and $\mathcal{H}$ is separable, then $\mathfrak{H}$ is separable, too. This is the case when $\mathfrak{X}$ is the Lebesgue-space.

If $\mathfrak{H}$ is separable, then it is the direct integral of $\{\mathcal{H}\}_{x \in X}$ over $\mathfrak{X}$ in the sense of Definition 14.1.1 of Kadison and Ringrose (1986):

$$\mathfrak{H} = \int_X^{\oplus} \mathcal{H}\, d\mu.$$

For the rest part of this section, we assume that $\mathfrak{H}$ is separable.

Definition A.7 An operator $T \in \mathcal{B}(\mathfrak{H})$ is said to be *decomposable* if there are operators $T_x \in \mathcal{B}(\mathcal{H})$ for $x \in X$ such that for each $a \in \mathfrak{H}$ we have

$$(Ta)(x) = T_x a(x) \quad \text{for almost all } x \in X$$

In this case, the system $\langle T_x : x \in X \rangle$ is called a *decomposition* of T and we write

$$T = \langle T_x : x \in X \rangle$$

Components of the decomposition are almost everywhere unique.

In particular, the identity operator $I_{\mathfrak{H}} \in \mathcal{P}(\mathfrak{H})$ is decomposable with the decomposition

$$I_{\mathfrak{H}} = \langle I_{\mathcal{H}} : x \in X \rangle$$

In a similar manner we can define for each $T \in \mathcal{B}(\mathcal{H})$ an operator $h(T)$ as follows:

$$h(T) = \langle T : x \in X \rangle$$

that is, for $a \in \mathfrak{H}$ the action of $h(T)$ is defined as

$$(h(T)a)(x) \doteq Ta(x) \quad \text{for all } x \in X$$

It is not hard to see that $h(T) \in \mathcal{B}(\mathfrak{H})$ for all $T \in \mathcal{B}(\mathcal{H})$.

By Theorem 14.1.10 of Kadison and Ringrose (1986), the set of decomposable operators in $\mathcal{B}(\mathfrak{H})$ is a von Neumann algebra acting on $\mathfrak{H}$. Define $\mathcal{N}_{obj+subj}$ to be the decomposable operators in $\mathcal{B}(\mathfrak{H})$. Now the map $h : \mathcal{B}(\mathcal{H}) \hookrightarrow \mathcal{N}_{obj+subj}$ is a $*$-algebra embedding with $h[\mathcal{P}(\mathcal{H})] \subseteq \mathcal{P}(\mathcal{N}_{obj+subj})$ and, in particular, as a function $h : \mathcal{P}(\mathcal{N}_{obj}) \to \mathcal{P}(\mathcal{N}_{obj+subj})$ it is a lattice embedding preserving orthocomplementation, as well.

We define the normal state $\phi_{obj+subj}$ of $\mathcal{N}_{obj+subj}$ as follows:

$$\phi_{obj+subj}(T) \doteq \int_X \phi_{subj}(T_x)\, d\mu(x) \quad \text{for all } T = \langle T_x : x \in X \rangle \in \mathcal{N}_{obj+subj}$$

Let $\mathcal{N}_{subj}$ be the von Neumann subalgebra

$$\mathcal{N}_{subj} = \{ r \cdot I_{\mathfrak{H}} : r \in \mathbb{R} \}$$

Next, let us define the conditional expectation $\mathbb{E} : \mathcal{N}_{obj+subj} \to \mathcal{N}_{subj}$ by the equation

$$\mathbb{E}(T) = \langle \phi_{obj}(T_x) \cdot I_{\mathcal{H}} : x \in X \rangle \quad \text{for all } T = \langle T_x : x \in X \rangle \in \mathcal{N}_{obj+subj}$$

$\mathbb{E}$ is indeed a conditional expectation: $\mathbb{E}(I_{\mathfrak{H}}) = I_{\mathfrak{H}}$ and if R_1, R_2 are in $\mathcal{N}_{subj}$, that is $R_1 = r_1 I_{\mathfrak{H}}$ and $R_2 = r_2 I_{\mathfrak{H}}$, then for all $T = \langle T_x : x \in X \rangle \in \mathcal{N}_{obj+subj}$ we have

$$\begin{aligned}
\mathbb{E}(R_1 T R_2) &= \langle \phi_{obj}(r_1 I_{\mathcal{H}}) \phi_{obj}(T_x) I_{\mathcal{H}} \phi_{obj}(r_2 I_{\mathcal{H}}) : x \in X \rangle \\
&= \langle r_1 \phi_{obj}(T_x) I_{\mathcal{H}} r_2 : x \in X \rangle \\
&= r_1 \mathbb{E}(T) r_2 = r_1 I_{\mathfrak{H}} \mathbb{E}(T) r_2 I_{\mathfrak{H}} \\
&= R_1 \mathbb{E}(T) R_2
\end{aligned}$$

$\mathbb{E}$ is also normal, that is $\mathbb{E}(\sup_\alpha X_\alpha) = \sup_\alpha \mathbb{E}(X_\alpha)$ holds for any sequence of uniformly bounded self-adjoint operators X_α. For, take any such sequence $X_\alpha = \langle X_{\alpha,x} : x \in X \rangle \in \mathcal{N}_{obj+subj}$ and use the equation $\sup_\alpha X_\alpha = \langle \sup_\alpha X_{\alpha,x} : x \in X \rangle$ and normality of ϕ_{obj} to obtain

$$\begin{aligned}
\mathbb{E}(\sup_\alpha X_\alpha) &= \langle \phi_{obj}(\sup_\alpha X_\alpha) \cdot I_{\mathcal{H}} : x \in X \rangle \\
&= \langle \sup_\alpha \phi_{obj}(X_\alpha) \cdot I_{\mathcal{H}} : x \in X \rangle \\
&= \sup_\alpha \mathbb{E}(X_\alpha)
\end{aligned}$$

Finally, let us check items (i)–(iii) and (A)–(B) in the statement. As for (i) we have already seen that $h : \mathcal{N}_{obj} \to \mathcal{N}_{obj+subj}$ is a von Neumann algebra embedding. $h(\mathcal{N}_{obj}) \cap \mathcal{N}_{subj} = \{0, I\}$ is clear from the definition of $\mathcal{N}_{subj}$. Item (ii) is straightforward from the definition of $\mathbb{E}$. As for item (iii): $\phi_{obj+subj}$ is a state defined in $\mathcal{N}_{obj+subj}$, therefore it remained to show properties (A) and (B).

(A) State extension: The restriction of $\phi_{obj+subj}$ to $h(\mathcal{N}_{obj})$ coincides with ϕ_{subj}. Indeed, for $\langle T : x \in X \rangle \in h(\mathcal{N}_{obj})$ we have by definition

$$\phi_{obj+subj}(\langle T : x \in X \rangle) = \int_X \phi_{subj}(T)\, d\mu(x) = \phi_{subj}(T)$$

(B) Principle: The restriction to $h(\mathcal{N}_{obj})$ of the $\mathbb{E}$-conditioned state $(\phi_{obj+subj}) \circ \mathbb{E}$
coincides with ϕ_{obj}. Indeed, given any $\langle T : x \in X \rangle \in h(\mathcal{N}_{obj})$ we have

$$\phi_{obj+subj}(\mathbb{E}(\langle T : x \in X\rangle)) = \int_X \phi_{subj}(\phi_{obj}(T) \times I) \, d\mu(x)$$

$$= \phi_{obj}(T) \times \int_X \phi_{subj}(I) \, d\mu(x) = \phi_{obj}(T)$$

$\square$

References

Bana, G. 2016. On the formal consistency of the principal principle. *Philosophy of Science 83*(5): 988–1001. University of Chicago Press.

Black, R. 1998. Chance, credence, and the principal principle. *The British Journal for the Philosophy of Science* 49: 371–385.

Briggs, R. 2009. The anatomy of the big, bad bug. *Noûs* 43: 428–449.

Bub, J., and M. Radner. 1968. Miller's paradox of information. *The British Journal for the Philosophy of Science* 19: 63–67.

Caves, C.M., C.A. Fuchs, and R. Schack. 2002. Quantum probabilities as Bayesian probabilities. *Physical Review A* 65: 022305.

Eberhardt, F., and C. Glymour. 2009. Hans Reichenbach's probability logic. In *Handbook of the history of logic*, ed. D.M. Gabbay, S. Hartmann, and J. Woods, vol. 10. Inductive logic. Amsterdam: Elsevier.

Fonseca, I., and G. Leoni. 2007. *Modern methods in the calculus of variations: L^p spaces*. New York: Springer.

Frigg R., and C. Hoefer. 2015. The best Humean system for statistical mechanics. *Erkenntnis* 80: 551–574.

Gaifman, H. 1988. A theory of higher order probabilities. In *Causation, chance, and credence. Proceedings of the Irvine conference on probability and causation, Volume 1*, ed. B. Skyrms, and W.L. Harper. The University of Western Ontario Series in Philosophy of Science, vol. 41, 191–219. Dordrecht: Kluwer Academic.

Glynn. L. 2010. Deterministic chance. *The British Journal for the Philosophy of Science* 61: 51–80.

Gyenis, Z., and M. Rédei. 2011. Characterizing common cause closed probability spaces. *Philosophy of Science* 78: 393–409.

Gyenis, Z., and M. Rédei. 2016. Measure theoretic analysis of consistency of the principal principle. *Philosophy of Science 83*(5): 972–987. University of Chicago Press.

Hall, N. 1994. Correcting the guide to objective chance. *Mind* 103: 505–518.

Hall, N. 2004. Two mistakes about credence and chance. *Australasian Journal of Philosophy* 82: 93–111.

Hoefer, C. 2007. The third way on objective probability: A sceptic's guide to objective chance. *Mind* 116: 449–596.

Howson, C., and G. Oddie. 1979. Miller's so-called paradox of information. *The British Journal for the Philosophy of Science* 30: 253–278.

Ismael, J. 2008. Raid! correcting the big bad bug. *Noûs* 42: 292–307.

Kadison, R.V., and J.R. Ringrose. 1986. *Fundamentals of the theory of operator algebras, Vol. I. and II*. Orlando: Academic Press.

Leifer, M.S., and R.W. Spekkens. 2013. Towards a formulation of quantum theory as a causally neutral theory of Bayesian inference. *Physical Review A* 88: 052130.

Lewis, D. 1980. A subjectivist's guide to objective chance. In *Studies in inductive logic and probability*, ed. R.C. Jeffrey, vol. II, 263–293. Berkely: University of California Press. Reprinted in D. Lewis, Philosophical papers, vol. II, Oxford University Press, 1986.

Lewis, D. 1986. A subjectivist's guide to objective chance. In *Philosophical papers*, vol. II, 83–132. Oxford: Oxford University Press.

Lewis, D. 1994. Humean supervenience debugged. *Mind* 103: 473–490.

Loewer, B. 2004. David Lewis' Humean theory of objective chance. *Philosophy of Science* 71: 115–1125.

Mackie, J. 1966. Miller's so-called paradox of information. *The British Journal for the Philosophy of Science* 14: 144–147.

Meacham, C.J.G. 2010. Two mistakes regarding the Principal Principle. *The British Journal for the Philosophy of Science* 61: 407–431.

Miller, D. 1966. A paradox of information. *The British Journal for the Philosophy of Science* 17: 59–61.

Nissan-Rozen, I. 2013. Jeffrey conditionalization, the Principal Principle, the desire as belief thesis and Adam's thesis. *The British Journal for the Philosophy of Science* 64: 837–850.

Pettigrew, R. 2012. Accuracy, chance and the principal principle. *Philosophical Review* 121: 241–275.

Pettigrew, R. 2013. A new epistemic utility argument for the principal principle. *Episteme* 10: 19–35.

Rédei, M. 1992. When can non-commutative statistical inference be Bayesian? *International Studies in the Philosophy of Science* 6: 129–132.

Rédei, M. 1998. *Quantum logic in algebraic approach*, vol. 91 Fundamental theories of physics. Dordrecht: Kluwer Academic Publisher.

Rédei, M., and S.J. Summers. 2007. Quantum probability theory. *Studies in History and Philosophy of Modern Physics* 38: 390–417.

Roberts, J.T. 2001. Undermining undermined: Why Humean supervenience never needed to be debugged. *Philosophy of Science* 68: S98–S108. Proceedings of the 2000 Biennial meeting of the philosophy of science association. Part I contributed papers.

Takesaki, M. 2003. *Theory of operator algebras Vols. I and II*. Encyclopedia of mathematical sciences, vol. 125. Berlin/Heidelberg: Springer.

Thau, M. 1994. Undermining and admissibility. *Mind* 103: 491–504.

Valente, G. 2007. Is there a stability problem for Bayesian noncommutative probabilities? *Studies in History and Philosophy of Modern Physics* 38: 832–843.

Vranas, P.B.M. 2004. Have your cake and eat it too: The old principal principle reconciled with the new. *Philosophy and Phenomenological Research* LXIX: 368–382.

Chapter 2
Is it the Principal Principle that Implies the Principle of Indifference?

Balázs Gyenis and Leszek Wroński

Abstract Hawthorne et al. (Br J Philos Sci, http://bjps.oxfordjournals.org/lookup/doi/10.1093/bjps/axv030) argue that the Principal Principle implies a version of the Principle of Indifference. We show that what the Authors take to be the Principle of Indifference can be obtained without invoking anything which would seem to be related to the Principal Principle. In the Appendix we also discuss several Conditions proposed in the same paper.

The argument of Hawthorne et al. (2015) ("Authors" from here on) that the Principal Principle implies what they take to be the Principle of Indifference is based on their Proposition 2, which (using their notation, and remembering that X "says that the chance at time t of proposition A is x and E is any proposition that is compatible with X and admissible at time t," ibid. p. 1.[1]) can be restated as follows:

Proposition 2 *Let* $P(A|X) = P(A|XE) = P(A|FXE) = P(A|(A \leftrightarrow F)XE)$. *Then from* $x \neq 0$, $x \neq 1$ *and from the Principal Principle* $P(A|X) = x$ *it follows that* $P(F|XE) = 1/2$.

Unfortunately it is unclear what the essence of Proposition 2 has to do with the Principal Principle. One can easily show that its "indifference" message is simply a consequence of imposing a sufficient number of independence constraints on an arbitrary probability space:

---

[1]We note here that the Authors require also that F be "contingent and atomic". From their footnote 1 it is clear that by this they do not mean that F should be an atom of the considered algebra of propositions, and rather that it should not be "logically complex". We believe that the mathematics of the current paper, and specifically the Counterexample in the Appendix is acceptable from that point of view.

B. Gyenis
Institute of Philosophy, Research Centre for the Humanities, Hungarian Academy of Sciences, Budapest, Hungary

L. Wroński (✉)
Jagiellonian University, Kraków, Poland
e-mail: leszek.wronski@uj.edu.pl

© Springer International Publishing AG 2017
G. Hofer-Szabó, L. Wroński (eds.), *Making it Formally Explicit*, European Studies in Philosophy of Science 6, DOI 10.1007/978-3-319-55486-0_2

Proposition 2' *Let* $P(A|X) = P(A|XE) = P(A|FXE) = P(A|(A \leftrightarrow F)XE)$. *Then from* $P(A|X) \neq 0$, $P(A|X) \neq 1$ *it follows that* $P(F|XE) = 1/2$.

The proof is a straightforward calculation which does not make use of the Principal Principle. As it can be readily seen by comparing the Authors' Proposition 2 with Proposition 2' the only tenuous connection between the Authors' Proposition 2 and the Principal Principle comes from that in order to conclude that $P(F|EX) = 1/2$ one needs to rule out the extreme cases when $P(A|X) = 0$ or $P(A|X) = 1$; Proposition 2 achieves this by setting $x \neq 0$, $x \neq 1$ and applying the Principal Principle, but this is an overkill, since $P(A|X) \neq 0$, $P(A|X) \neq 1$ could be directly achieved by simply requiring that learning event X does not make A impossible or certain.

Thus the Principal Principle does not have much to do with the essence of Proposition 2. The Authors themselves also make a side-remark on p. 4. that "interestingly, [their] line of argument does not depend on the structure of the proposition X," but they do not follow up on it and they do not seem to come to the realization that the indifference message of Proposition 2 is a mere consequence of the independence constraints: their Superprincipal Principle, with which the Authors generalize the Principal Principle, also makes assumptions about the structure of proposition X, and their generalized argument given subsequently still relies on equating a conditional probability $P(A|XE)$ with a number $x \neq 0$, $x \neq 1$ to reach the conclusion. Assuming that $P(A|XE) = x$ and $x \neq 0$, $x \neq 1$ is still an overkill to merely ensure $P(A|XE) \neq 0$, $P(A|XE) \neq 1$, which as we can see in Proposition 2' is all that is needed to arrive at $P(F|XE) = 1/2$. And thus the Superprincipal principle does not have much to do with the essence of Proposition 2 either.

Notice that our claim is not simply that it is possible to obtain the Principle of Indifference from something weaker than the Principal Principle: this would even strengthen the original point the Authors wanted to make! We say rather that the argument does not seem to have any connection to the Principal Principle, since it is an inference from a set of independence assumptions to a conclusion about independence, not relying on any statement regarding (credences about) chances.

The Authors devote some time to "defending" their Condition 2 which is intended to furnish one of the premises of their Proposition 2. It can be restated as follows:

Condition 2 If $P(A|XE) = P(A|X)$, $P(F) = P(F|XE)$, and the Principal Principle $P(A|X) = x$ holds, then $P(A|XE(A \leftrightarrow F)) = P(A|X)$.

The Authors state that Condition 2 "must hold because [it] encapsulate[s] core intuitions about defeat" (p. 4.); however, one can show with a counterexample that Condition 2 is not a theorem of probability theory. Instead of searching for ways to "defend" Condition 2 it might be instructive to ask what additional supposition (S) the Authors would need to add to the premises of Condition 2 to make it a theorem. It is easy to find candidates for (S) that express probabilistic requirements in terms of A, E, F, X (and possibly other events) without referring to the value x; however if (S) took such form, then the same (S) could also be added to the premises of

Condition 2' If $P(A|XE) = P(A|X)$, $P(F) = P(F|XE)$ holds, then $P(A|XE(A \leftrightarrow F)) = P(A|X)$

to make this latter a theorem. This is since if (S) does not refer to x, then the additional assumption $P(A|X) = x$, which is what differentiates Condition 2 from Condition 2', will clearly not be used in the derivation. Thus if the Authors succeed in finding and defending such an (S) with which Condition 2 furnishes the desired premise of Proposition 2, then with the same (S) Condition 2' also furnishes the corresponding premise of Proposition 2'. Since neither Condition 2' nor Proposition 2' invokes the Principal Principle, we again would complete an argument for what the Authors' interpret as the Principle of Indifference without invoking the Principal Principle. (For the mentioned counterexample, proof of Proposition 2', and additional comments on the Authors' arguments regarding their Conditions 1 and 2 the Reader is referred to the Appendix.)

Acknowledgements Balázs Gyenis was supported by the National Research, Development and Innovation Office, K 115593 grant. Leszek Wroński was supported by Tomasz Placek's MISTRZ grant "Probability, modalities and Bell theorems" of the Foundation for Polish Science. The work on the current paper was carried out as a part of the "Probability, Causality and Determinism" Bilateral Mobility Grant of the Hungarian and Polish Academies of Sciences, NM-104/2014.

The authors would like to thank Gergei Bana and Christian Wallmann for insightful comments on a draft of this paper.

Appendix

Proof of Proposition 2'

Before the proof for later purposes we slightly reformulate Proposition 2':

Proposition 2' *Let $(\mathcal{L}_X, P_X)$ be a probability space, $A, E, F \in \mathcal{L}_X$, and let $P_X(A) = P_X(A|E) = P_X(A|FE) = P_X(A|(A \leftrightarrow F)E)$. Then from $P_X(A) \neq 0$, $P_X(A) \neq 1$ it follows that $P_X(F|E) = 1/2$.*

Proof Using the following identification for the probabilities for different conjunctions:

	o_1	o_2	o_3	o_4	o_5	o_6	o_7	o_8
$P_X(.)$	p_1	p_2	p_3	p_4	p_5	p_6	p_7	p_8
A					x	x	x	x
E			x	x			x	x
F		x		x		x		x

(so that $AEF = o_8$, $P_X(AEF) = p_8$ etc.) we can restate the conditions as

$$p_8/(p_4 + p_8) = p_8/(p_3 + p_8) \tag{2.1}$$

$$p_8/(p_4 + p_8) = p_5 + p_6 + p_7 + p_8 \tag{2.2}$$

$$p_8/(p_3 + p_8) = (p_7 + p_8)/(p_3 + p_4 + p_7 + p_8); \tag{2.3}$$

we also require $p_i \geq 0$, $\sum_{i=1}^{8} p_i = 1$, and $p_3 + p_4 + p_7 + p_8 > 0$, $p_4 + p_8 > 0$, $p_3 + p_8 > 0$ for the conditional probabilities to make sense.

By (2.1) we get that either $p_3 = p_4$ or $p_8 = 0$. If $p_8 = 0$ then from (2.2) $P_X(A) = p_5 + p_6 + p_7 + p_8 = 0$ and all other conditions can be satisfied. If $p_8 \neq 0$ then assuming $p_3 = p_4$ from (2.3) we get that either $p_7 = p_8$ or $p_4 = 0$. If $p_4 = 0$ then from (2.2) $P_X(A) = p_5 + p_6 + p_7 + p_8 = 1$ and all other conditions can be satisfied. Assuming $p_8 \neq 0$, $p_4 \neq 0$ and thus that $p_3 = p_4$ and $p_7 = p8$, $P_X(F|E) = (p_4 + p_8)/(p_3 + p_4 + p_7 + p_8) = (p_4 + p_8)/(2p_4 + 2p_8) = 1/2$ and all other conditions can be satisfied. To sum up, either $P_X(A) = 0$, or $P_X(A) = 1$, or $P_X(F|E) = 1/2$. **QED**

Remarks on Conditions 1 and 2 and a Counterexample

The premises of the Authors' Proposition 2 are supposed to be furnished by the following two Conditions:

"Condition 1. If E is not a defeater and XE contains no information that renders F relevant to A, then EF is not a defeater."

"Condition 2. If E is not a defeater and XE contains no information relevant to F, then $E(A \leftrightarrow F)$ is not a defeater." (ibid., p. 2)

which are, as mentioned earlier, to be understood in a context where X says that the chance at time t of proposition A is x. The Authors call E a "non-defeater" (w.r.t. A) if $P(A|XE) = P(A|X) = x$ holds.

Unfortunately the Authors "take the supposition that XE contains no information that renders F relevant to A to imply that $P(A|FXE) = P(A|XE)$" (ibid., p. 2). In other words, according to the Authors, if XE contains no information that renders F relevant to A then XE screens off F from A. This is surely too strong. The most natural mathematical counterpart of the Author's natural language expression in Condition 1 seems to be the one corresponding closely to "if F is not relevant to A then XE does not make it relevant", namely

$$P(A|F) = P(A) \rightarrow P(A|FXE) = P(A|XE). \tag{2.4}$$

With that in place, Condition 1 becomes formally the following:

Condition 1 (reformulated) If $P(A|XE) = P(A|X)$, $P(A|F) = P(A) \rightarrow P(A|FXE) = P(A|XE)$, and the Principal Principle $P(A|X) = x$ holds, then $P(A|FXE) = P(A|X)$.

Formula (2.4) is indeed weaker than the Authors' screening off condition, as evidenced by the following Counterexample (which—we claim—falsifies also the original Conditions 1 and 2):

Counterexample 1 *Let $\mathcal{L}$ be the Boolean algebra generated by the logically independent propositions $A, E, F,$[2] and $X = $ "$Ch(A) = 2/5$" and let P be the probability on $\mathcal{L}$ that assigns $1/10$ to the propositions $\neg X \neg A \neg E \neg F$, $\neg X \neg AE \neg F$, $X \neg A \neg E \neg F$, $X \neg AE \neg F$, and assigns $1/20$ to the rest of the algebraic atoms. $(\mathcal{L}, P)$ then has the following properties:*

$$\frac{2}{5} = P(A|XE) = P(A|X) = \frac{2}{5} \tag{2.5}$$

$$\frac{1}{2} = P(A|F) \neq P(A) = \frac{2}{5} \tag{2.6}$$

$$\frac{1}{2} = P(A|FXE) \neq P(A|XE) = \frac{2}{5} \tag{2.7}$$

$$\frac{1}{2} = P(A|FXE) \neq P(A|X) = \frac{2}{5} \tag{2.8}$$

$$\frac{2}{5} = P(F) = P(F|XE) = \frac{2}{5} \tag{2.9}$$

$$\frac{1}{3} = P(A|XE(A \leftrightarrow F)) \neq P(A|X) = \frac{2}{5} \tag{2.10}$$

(i) Due to (2.5), (2.6), and (2.9) $(\mathcal{L}, P)$ is a counterexample to

$$P(A|XE) = P(A|X) = x$$
$$\frac{P(A|F) = P(A) \rightarrow P(A|FXE) = P(A|XE)}{P(A|FXE) = P(A|X) = x,}$$

that is, to the reformulated Condition 1 (see below)

---

[2] And so F can be assumed not to be "logically complex", as the Authors wish.

(ii) *Due to (2.5), (2.6), (2.7), and (2.8) $(\mathcal{L}, P)$ is a counterexample to Condition 1. (On the assumption that from the fact that due to (2.6) and (2.7) we see that conditioning on XE leaves both the probability of A and the probability of F intact we can infer that "XE contains no information that renders F relevant to A".)*

(iii) *Due to (2.5), (2.9), and (2.10) $(\mathcal{L}, P)$ is a counterexample to Condition 2.*

The interpretation of the propositions involved has to matter since the Authors only provide a single intuition boosting example to illustrate the reasonableness of Condition 1, and the plausibility of their illustration rests on a proper reading of what it is for a characterization of a situation to not contain information that renders two events relevant to each other. Apart from this sole example the Authors do not provide any philosophical analysis of the plausibility of Condition 1, because they claim, with their Proposition 1, that Condition 1 "provably holds." However, it holds only if "containing no information that renders F relevant to A" is expressed as "screening off F from A"; our Counterexample also shows that it does not hold if we follow the formulation of the condition more accurately, for example by using (2.4).

There are of course other formal candidates for "XE containing no information that renders F relevant to A" which could be considered, of various logical strength. One which perhaps could be promising is $P(A|F) - P(A) = P(A|FXE) - P(A|XE)$. However, if we use *that* in Condition 1, it agains falls prey to our Counterexample 1—as evidenced by the clause (ii). The situation is similar with all other candidates we have explored.

Since their Proposition 1 is no proof of Condition 1 as reconstructed above, the Authors would need to provide a philosophical justification for it. But since the justification for Condition 2 also depends on the availability of justification of Condition 1—the Authors counter one of the main objections against Condition 2, namely that it fails under a definition of admissibility put forward by Meacham (2010), by claiming that that Meacham's admissibility is too restrictive is evidenced by the fact that it entails a violation of the supposedly provable Condition 1—the premises of the main Proposition 2 stay insufficiently motivated, which then carries over to whatever conclusion one can draw from it.

To wager philosophical bets on the potential success of justification of Conditions 1 and 2 we remark that our Counterexample also evidences a probability space in which both of these conditions fail. Universal quantification (or at least the existence of a sound argument for typicality) over events and spaces matters since the mere fact that some events in some probability spaces have a probability $1/2$ would not establish a connection between the Principal Principle and the Principle of Indifference.

We emphasize that our remarks regarding the appropriate formal reconstruction of the Authors' verbal formulation of Condition 1 are independent from the argument developed in the main text.

As for some further remarks regarding Condition 2, we note that the flavor "under some conditions, since we are ignorant of whether an independence constraint holds, we assume that it does" it possesses is similar to "under some conditions, since we are ignorant of the probability, we assume that it is 1/2"—which is the Principle of Indifference. This raises the suspicion that we are proving what we already put in; indeed a requirement of probabilistic independence is of comparable strength as narrowing down the probability to a single value, as we are now going to illustrate.

Thus let us consider now the Authors' inference from $P(F|XE) = 1/2$ to $P(F) = 1/2$ right after their Proposition 2, which is the basis for their philosophical claim that the Principle of Indifference should also hold for the unconditional prior. It should be clear that this inference is only valid if Condition 2 indeed holds, but the assumptions of their Proposition 2 can hold without Condition 2 being true (or, alternatively, assumptions of Proposition 2' can hold without Condition 2' being true). One can also quantify how frequently this is the case. Let us assume that $(\mathcal{L}_X, P_X)$ satisfies the conditions of our Proposition 2' and hence $P_X(F|E) = 1/2$. Using the terminology of the proof of Proposition 2', let $\mathcal{L}$ be generated by adding an additional algebraic atom o_0 to $\mathcal{L}_X$, let A, E, F be the same as in the proof and let $X = \{o_1, \ldots, o_8\}$; then with the probability $P(Y) = p_0 \cdot \chi_Y(o_0) + (1 - p_0) \cdot P_X(YX)$ the new $(\mathcal{L}, P)$ also satisfies the conditions of Proposition 2' and $P(Y|X) = P_X(Y)$. It is then easy to show that $P(F) = 1/2$ only when $p_0 = 1 - 1/(2(p_2 + p_4 + p_6 + p_8))$. Thus even though in all of the so-defined extensions $P(F|EX) = 1/2$, there is only one in which $P(F) = 1/2$ (and thus in which the independence constraint $P(F) = P(F|XE)$ holds) while in all of the uncountably many others $P(F) \neq 1/2$. Loosely speaking it appears, then, that in this case we should expect Condition 2 to hold only as frequently as we would expect $P(F) = 1/2$ to hold if we set its value randomly.

References

Hawthorne, J., J. Landes, C. Wallmann, and J. Williamson. 2015. The Principal Principle implies the Principle of Indifference. *The British Journal for the Philosophy of Science*. http://bjps. oxfordjournals.org/lookup/doi/10.1093/bjps/axv030.

Meacham, C.J. 2010. Two mistakes regarding the Principal Principle. *British Journal for the Philosophy of Science* 61: 407–431.

Chapter 3
Models of Objective Chance: An Analysis through Examples

Gergei Bana

Abstract In his seminal work, A Subjectivist's Guide to Objective Chance, David Lewis considered the possibility for a subjectivist to accept the existence of objective chance, and argued how this acceptance would affect the subjectivist's degrees of belief (credences) about the world: they have to satisfy the Principal Principle. Lewis did not put his proposal into mathematically precise terms. Most importantly, he did not define what kind of mathematical object objective chance was. In this work we pay careful attention to identify what mathematical model the subjectivist agent considers, and through several simple examples we illustrate how objective chance can be incorporated in the subjectivist's model in a mathematically rigorous manner.

3.1 Introduction

In A Subjectivist's Guide to Objective Chance, David Lewis (1980) considered the possibility for a subjectivist to accept the existence of objective chance, and argued how this acceptance would affect the subjectivist's degrees of belief (credences) about the world. He asked what constraints the acceptance of the existence of objective chance imposed on the credences that a reasonable subjectivist could assign to sets of possible worlds. The starting point of Lewis (1980) is that we accept the existence of a time dependent chance ch_t, and that both objective chance and credence should satisfy the usual rules of probability theory. Lewis's answer to the above question is that the only limitation on the credence function (measure) is that for any chancy event A, conditioned on the evidence the agent has collected and on that the objective chance of A is known to be r, the credence of A also has to equal r:

$$C(A|ch_t(A) = r \wedge E) = r$$

G. Bana (✉)
Faculty of Science, Technology and Communication, University of Luxembourg,
6 Avenue de la Fonte, L 4365 Esch sur Alzette, Luxembourg
e-mail: bana@math.upenn.edu

© Springer International Publishing AG 2017
G. Hofer-Szabó, L. Wroński (eds.), *Making it Formally Explicit*, European Studies
in Philosophy of Science 6, DOI 10.1007/978-3-319-55486-0_3

as long as E is "admissible", that is, as long as the evidence collected does not force us to have a degree of belief other than r.

Lewis did not state his proposal in a mathematically rigorous manner. In the second part of his work, Lewis (1980) mentions that objective chance is a chance distribution $P_{tw}(A)$ that assigns a probability to event A, and it can depend on time t and on possible world w but does not investigates what properties this function should satisfy. To this day a lot of research is being done on the Principal Principle, especially around the mystery of what admissibility means. These works seem to us rather speculative as long as we do not specify mathematically clearly what we are talking about.

At the PSA Biennial Meeting in 2014, there were some attempts to treat at least some particular questions about the Principal Principle in a mathematically precise manner by Rédei and Gyenis (2016) and Bana (2016). While preparing the latter, I realized that a clear understanding of what objective chance is mathematically is entirely missing from the literature.

My aim here is to look at several specific examples, through which we can identify what kind of mathematical object corresponds to objective chance in its most general sense. I believe that the mathematical model we give is the most general one in which the notion of objective chance can be meaningful. We do not rely on special assumptions for our analysis, I try to be completely faithful to Lewis's original 1980 work, but make it more explicit.

In this work I do not attempt to treat the Principal Principle itself. The sole purpose here is to motivate a rigorous definition of objective chance, and I shall work out the consequences of this definition on the Principal Principle elsewhere.

Furthermore, in this work I do not attempt to formalize Lewis's later views (for example, Lewis 1994) on how objective chance is given by a best system. We simply propose a general modeling of objective chance that can accommodate any kind of objective chances, let it be propensity, relative frequencies or determined by a best system.

3.2 The Agent and His Model

We shall be careful throughout this work detaching the model of reality from reality itself. We do not follow Lewis in speculating what objective chance in reality is, as it does not matter. What matters is what the agent's view about reality is, what model of possible worlds the agent choses as a starting point in which he places the pieces of evidence he collects. *As Lewis's starting point is the assumption that objective chance is real, our main task is to see how to incorporate objective chance rigorously in the mathematical model of possible worlds.* We want to give a general mathematical definition that objective chance must minimally satisfy let it be a propensity or reflect relative frequencies.

Lewis did not make explicit how the subjectivist agent views reality, and how he collects evidence, so let us make this explicit here:

- **Subjecivist Agent's View:** The subjectivist agent has a set W, such that each $w \in W$ represents a possible world. In the agent's opinion, each $w \in W$ describes facts that are either true or not in the real world. The agent assumes that in this W there is a w_{real} corresponding to the real world, but he does not know which w it is. So, he assigns degrees of belief, C called credence to subsets $S \subseteq W$, and $C(S)$ represents his degree of belief that $w_{real} \in S$. Lewis assumes that C satisfies the laws of probability. Accordingly, on W a σ-algebra Σ of subsets is defied, and C is a probability distribution on Σ. We can call (W, Σ) the *subjectivist's model* of the world. The agent, one way or other, we do not care how, collects evidence. What we do care about is that whenever he collects a piece of evidence, he places that in his mathematical model, that is, he assigns a set $E \in \Sigma$ to it. Having collected such a piece of evidence E, from that on the agent assumes that $w_{real} \in E$, meaning that he updates his credence to $C'(\cdot) := C(\cdot \,|E)$. When the agent learns one piece of evidence after the other, $E_1, E_2, \ldots, E_n \in \Sigma$, then he updates his credence to $C_n(\cdot) := C(\cdot \,|E_1 \wedge E_2 \wedge \ldots \wedge E_n)$.

The set of possible worlds, W, may be produced in many ways. It may be a set of possible worlds that abide a certain theory of physics, it can be about coin toss, etc. One way or other, we do not care, it has to be defined mathematically precisely for the subjectivist agent to assign his credences.

E could be a set of W the elements of which are possible worlds corresponding to manufacturing a coin in a certain way, or to forming a certain radioactive material, or to a certain prediction of a crystal ball, etc. Normally, this happens the following way: An observable, such as the outcome of a coin toss, a radioactive decay, etc. is represented mathematically as a random variable (measurable function) X on W. To each possible world $w \in W$, a value $X(w)$ of the observable belongs. Measuring the value of the random variable, and obtaining say x, the set corresponding to this value, $E = \{w|X(w) = x\}$ is in Σ (as X is measurable), and this is an example of an evidence.

Note, this all is the agent's view! *It is entirely irrelevant for us whether the agent's model is a good model of the world or not. We only care about that he has a model and that pieces of evidence are placed in this model.*

Having accepted this modeling of the subjectivist agent's view, next question is how to accommodate objective chance in this model in a mathematically rigorous manner. For the subjectivist who believes in objective chance, objective chance has to be part of the model of possible worlds, hence it should be defined somehow on W. Namely, when the objective chance of event A, that is, $ch_t(A)$ can have different possible values in the agent's view, that means that each of these different possible values of $ch_t(A)$ must signify a different possible world. In other words, to each possible world w, a single value $ch_{tw}(A)$ must belong. Let us rather denote it by $ch_t(w, A)$. Can the function $(w, A) \mapsto ch_t(w, A)$ be an arbitrary one, or does it have to satisfy certain properties? Clearly, as objective chance is a probability, for fixed t and w, the function $A \mapsto ch_t(w, A)$ should be a probability measure over some algebra, as Lewis required it. The main objective of this work is to investigate whether we need to require some other properties as well for ch_t that Lewis did not notice. We shall see that a chancy situation comes with some structures on the set of possible worlds that Lewis failed to consider, and the above function is not arbitrary relative to these structures. A further question is what happens when we allow time to change as well, what restrictions do we get for the function $t \mapsto ch_t$. More generally, what are the requirements that would make a family of objective chances $\{ch_i\}_{i \in I}$ consistent? This is a non trivial question, as it is easy to give examples of contradictory objective chances. However, this question is out of the scope of this paper, we shall only focus on what makes a function $(w, A) \mapsto ch(w, A)$ an objective chance function.

Through the examples below, I attempt to make a convincing case for the mathematical definition I propose in Sect. 3.4.

3.3 Examples

In this section I provide numerous examples to support my proposition for modeling objective chance. First we look at examples in which objective chance is a propensity, while in the last example objective chance corresponds to relative frequencies.

3.3.1 The Basics

Our first example is straight out of Lewis (1980) "... suppose you are not sure that the coin is fair. You divide your belief among three alternative hypotheses about the chance of heads, as follows.

- You believe to degree 27% that the chance of heads is 50%.
- You believe to degree 22% that the chance of heads is 35%.
- You believe to degree 51% that the chance of heads is 80%.

Then to what degree should you believe that the coin falls heads? Answer. (27% × 50%) + (22% × 35%) + (51% × 80%); that is, 62%."

3.3.1.1 A Set of Possible Worlds

How can we treat this in a mathematically rigorous manner? Perhaps the simplest set of possible worlds that can accommodate this experiment is when first a coin is forged with three kinds of possible biasses, and then the coin is tossed with heads or tails as possible outcomes. Perhaps the simplest imaginable mathematical model for this is the following: The set of possible worlds have elements of the form of ordered pairs (x, y) where x is one of the symbols b_1, b_2 or b_2 corresponding to the three possible biasses, and y is either symbol h or symbol t corresponding to heads and tails:

$$W := \left\{ (x, y) \,\middle|\, (x = b_1 \lor x = b_2 \lor x = b_3) \land (y = h \lor y = t) \right\}.$$

Accordingly, the world in which the proposition "the coin is manufactured so that it has the second kind of bias and the outcome of the coin toss is heads" is satisfied corresponds to the mathematical object $(b_2, h) \in W$. Let B_i (for $i = 1, 2, 3$) be the proposition that "the coin was manufactured with the i'th kind of bias". The subset of W corresponding to this proposition is denoted by $[B_i]$, and is

$$[B_i] := \{(b_i, h), (b_i, t)\} \subset W.$$

The set corresponding to the proposition (denoted by H) that "the outcome of the coin toss is heads" is

$$[H] := \{(b_1, h), (b_2, h), (b_3, h)\} \subset W,$$

and similarly for tails, $[T]$.

We can also think of $[H]$ and $[T]$ the following way: the outcomes of the coin toss is in fact a random variable on W:

$$V : W \to \{h, t\}, \qquad V((x, y)) = y.$$

Then

$$[H] = V^{-1}(\{h\}) \qquad \text{and} \qquad [T] = V^{-1}(\{t\}).$$

3.3.1.2 The Event Algebra of Chancy Outcomes and Objective Chance

We still have to tell what the chance ch of $[H]$ and of $[T]$ are. How can we incorporate chance as a mathematical object in the model we have defined? Note that there is a subalgebra[1] Σ_{chancy} of Σ corresponding to the outcomes of the coin toss generated by $[H]$ and $[T]$:

$$\Sigma_{\text{chancy}} = \{ \emptyset, [H], [T], W \}$$

In other words, Σ_{chancy} is the algebra we obtain by pulling back the discrete measure of $\{h, t\}$ with V.

According to Lewis's example, if the coin was manufactured the first way, then the chance of heads is 0.5, if the second way, then the chance of heads is 0.35, and if the third way, then the chance of heads is 0.8. That is, there is no single $ch([H])$, it has three possible values. In other words, ch cannot be a measure on Σ_{chancy}. However, for each possible $w \in W$, we can define $ch(w, [H])$. And this is completely intuitive: On each possible world, $[H]$ has some objective chance, and

---

[1] That is, closed under finite union and complementation.

it of course can vary from possible world to possible world. In fact, Lewis thought of this too: $ch(w, A)$ is nothing but his *chance distribution* on page 276 of Lewis (1980).

We have the following:

$$ch((x, y), [H]) := \begin{cases} 0.5 \text{ if } x = b_1 \\ 0.35 \text{ if } x = b_2 \\ 0.8 \text{ if } x = b_3 \end{cases}$$

Then, setting

$$ch((x, y), [T]) := 1 - ch((x, y), [H]), \quad ch((x, y), \emptyset) := 0, \quad ch((x, y), W) := 1$$

for each $w \in W$, the function

$$ch(w, \cdot) : A \mapsto ch(w, A)$$

becomes a probability measure on Σ_{chancy}, corresponding to the fact that on each possible world, the objective chance is a probability distribution on the possible outcomes.

Note that for all $A \in \Sigma_{\text{chancy}}$, the function

$$ch(\cdot, A) : w \mapsto ch(w, A)$$

is constant on the sets $[B_1]$, $[B_2]$, and $[B_3]$. This corresponds to the fact that the objective chance depends only on how the coin was manufactured. In other words, $ch(\cdot, A)$ *depends only on the past* before the coin is tossed. The set corresponding to the proposition $ch(\cdot, A) = r$, notated as $[ch(\cdot, A) = r]$ is $[B_1]$ if $r = 0.5$, $[B_2]$ if $r = 0.35$, and $[B_3]$ if $r = 0.8$, otherwise it is the emptyset.

3.3.1.3 The Event Algebra of the Probabilistic Setup and Historic Evidence

What is *historic evidence* in our example at the point when the coin is tossed? Clearly, statements about how the coin was manufactured, what kind of setup was carried out before the coin is tossed. Hence mathematically, in this example, historic evidence is an element in the subalgebra of events $\Sigma_1 \subset \Sigma$ generated by $[B_1]$, $[B_2]$, and $[B_3]$ as atoms. Written explicitly,

$$\Sigma_1 := \{ \emptyset, [B_1], [B_2], [B_3], [B_1] \cup [B_2], [B_2] \cup [B_3], [B_1] \cup [B_3], W \}.$$

For those who are familiar theory of stochastic processes, setting $\Sigma_0 := \{ \emptyset \}$ and $\Sigma_2 := \Sigma$, the sequence $\Sigma_0, \Sigma_1, \Sigma_2$ is a *filtration*: an assignment of increasing event algebras to time steps. The index i is time, it means "step i": 0 before the process starts, 1 for the forging of the coin and 2 two for the coin toss. Functions that are measurable[2] with respect to Σ_i are those that can be computed based on the information up until the i'th step is carried out. For example, a function that is measurable with respect to Σ_1 can depend on x, but not on y in $(x, y) \in W$ as by step 1 the coin has been forged, but not tossed.

We introduce the notation Σ_{env}, which shall stand for the chancy environment, in this example

$$\Sigma_{\mathrm{env}} := \Sigma_1$$

meaning that the chancy events are considered in the environment when the past until step 1 is fixed.

The property that $ch(\cdot, A)$ is constant on the sets $[B_1]$, $[B_2]$, and $[B_3]$ (the atoms of Σ_{env}) can be equivalently reformulated as $ch(\cdot, A)$ is measurable with respect to $\Sigma_{\mathrm{env}} = \Sigma_1$. As a consequence, an explicit hypothesis about the objective chance, $[ch(\cdot, A) \in I]$, where $I \subseteq [0, 1]$ an interval, is in Σ_1, hence giving such an explicit hypothesis is a special way of giving historical information in this example.

Note that for example,

$$ch((b_2, h), [H]) = ch((b_2, t), [H]) = 0.35,$$

even though $(b_2, t) \notin [H]$. This corresponds to the fact that although on (b_2, t), once the coin was manufactured to have b_2 bias, the chance of heads is 0.35, but it does not actually come about, because on this possible world the coin lands tails.

3.3.1.4 Extending the Chancy Algebra

If desired, for each w, the chance of any event in Σ can be defined relative to Σ_{env}. Let $\overline{ch}$ denote the extended chance. Now $\overline{ch}(\cdot, S)$ should be measurable with respect to Σ_{env} for all $S \in \Sigma$ corresponding to the idea that we are considering objective chance once the coin was manufactured with some bias. We can do this the following way. First let us ask the question: How could we define, say, $\overline{ch}((b_1, t), [B_2])$? Clearly, once the coin was manufactured to have bias b_1, that is to be fair, it can land heads or tails with probability 0.5, but it cannot change its bias. So $\overline{ch}((b_1, t), [B_2]) = 0$. For the same reason, $\overline{ch}((b_1, t), [B_1]) = 1$. In general, we set for all $E \in \Sigma_{\mathrm{env}}$,

---

[2]Recall that a function $f : W \to \mathbb{R}$ is measurable with respect to Σ σ-algebra on W if and only if for all intervals $I \subseteq \mathbb{R}$, the set $f^{-1}(I) := \{w \in W : f(w) \in I\} \in \Sigma$. In case Σ is atomic (finite spaces are always such), measurable functions are those that are constant on the atoms of Σ.

$$\overline{ch}(w, E) := \begin{cases} 1 \text{ if } w \in E \\ 0 \text{ if } w \notin E \end{cases}$$

and then for any $A \in \Sigma_{chancy}$ and $E \in \Sigma_{env}$,

$$\overline{ch}(w, E \cap A) := \overline{ch}(w, E) \cdot ch(w, A).$$

As all elements in Σ can be written as the union of sets of the form $E \cap A$ as above (the event algebra Σ is generated by Σ_{env} and Σ_{chancy}), this defines a unique probability measure $S \mapsto \overline{ch}(w, S)$ on Σ, and the function $w \mapsto \overline{ch}(w, S)$ is Σ_{env}-measurable for all $S \in \Sigma$.

3.3.1.5   Observations on the Structure

To summarize, we have seen in this example that to model the chancy situation, we can define a subalgebra Σ_{chancy} representing the chancy events of outcomes, and another subalgebra Σ_{env} representing the events about how the chancy experiment is set up, that is, how the coin is forged. Furthermore, we had a chance assignment $ch : W \times \Sigma_{chancy} \rightarrow [0, 1]$ such that for all $w \in W$, the function $ch(w, \cdot)$ is a probability measure on Σ_{chancy}, and for each $A \in \Sigma_{chancy}$, the function $ch(\cdot, A)$ is measurable with respect to Σ_{env}. These structures, Σ_{env} and Σ_{chancy} are both missing from Lewis's analysis and from the followup literature as well, but I think they are essential aspects of objective chance, and no rigorous analysis is possible without them.

3.3.2   What Is Time?

Let us continue by making our first example more complex. Suppose that the coin forging process is the following: It is first decided in some unknown way whether the coin to be forged shall be fair or not. If it is not going to be fair, then it is decided in a fair manner—for example, by throwing an old fair coin—whether the new coin shall be forged such that the bias allows a chance of 0.35 for heads, or a chance of 0.8. Once the coin is forged, it is thrown 3 times.

What is the new W? For example, the following. Similarly to the first example, let the symbol b_1 stand for the forging of the new coin in the first, fair way. Let a stand for forging the new coin biassed. Let b_2 and b_3 denote the forging of the coin the second and the third way again. Let t_1, t_2, t_3 and h_1, h_2, h_3 stand for the outcomes of the coin tosses. Clearly, forging the biassed coins is a longer process because it involves an extra decision. Let

$$W := W_1 \cup W_2$$

where

$$W_1 := [B_1] := \left\{ (x, y_1, y_2, y_3) \mid x = b_1, \ (y_i = t_i) \vee (y_i = h_i) \right\}$$

and

$$W_2 := \left\{ (w, x, y_1, y_2, y_3) \mid w = a, \ (x = b_2 \vee x = b_3), \ (y_i = t_i) \vee (y_i = h_i) \right\}$$

Let π_i denote the projection on the i'th entry. For example, $\pi_2((w, x, y_1, y_2, y_3)) = x$ and $\pi_2((x, y_1, y_2, y_3)) = y_1$. For a set $S \in W$, let

$$\pi_i(S) := \{ z \mid \text{there is a } w \in W \text{ such that } z = \pi_i(w) \}.$$

We take again $\Sigma := \mathcal{P}(W)$, that is, all subsets of W. Here too we have a filtration: $\Sigma_0, \Sigma_1, \Sigma_2, \Sigma_3, \Sigma_4, \Sigma_5$ where $\Sigma_0 := \emptyset$, $\Sigma_5 := \Sigma$, and for $j = 1, 2, 3, 4$, Σ_j is generated by the sets of the form $S \in \Sigma$ such that $\pi_1(S), \ldots, \pi_j(S)$ each has a single element, and S contains all elements of W that start with $\pi_1(S), \ldots, \pi_j(S)$:

$$\Sigma_j = \overline{\left\{ S \in \Sigma \ \middle| \ \begin{array}{l} \text{each of } \pi_1(S), \ldots, \pi_j(S) \text{ contains exactly one element,} \\ \text{and for all } w \in W, \text{ if } w \text{ is such that} \\ \pi_1(w) \in \pi_1(S) \wedge \ldots \wedge \pi_j(w) \in \pi_j(S), \text{ then } w \in S \end{array} \right\}}$$

where the overline means closure under making unions and complementation. That is, Σ_j is generated by sets S such that all the elements of each S have the same history until the j'th step and their future after j is completely undetermined.

For example, the set $\{(b_1, t_1, t_2, t_3), (b_1, t_1, t_2, h_3)\} \in \Sigma_3$ is such a set where the history until the third step is determined to be b_1, t_1, t_2, and after the third step it is completely open (t_3 or h_3). Or, $\{(a, b_2, t_1, h_2, h_3), (a, b_2, t_1, h_2, t_3), (a, b_2, t_1, t_2, h_3), (a, b_2, t_1, t_2, t_3)\} \in \Sigma_3$ is also such a set, where the history until the third step is fixed to be a, b_2, t_1, while the last two steps are completely open and can be h_2, h_3, or t_2, h_3, or h_2, t_3, or t_2, t_3. On the other hand, $\{(a, b_2, t_1, t_2, h_3), (a, b_2, t_1, h_2, t_3)\}$ is not such a set and is not in Σ_3, because although the history until the third step is fixed to be a, b, t_1, the last two are not completely undetermined as only t_2, h_3 and h_2, t_3 appear in the set. A function $f : W \to \mathbb{R}$ that has different values on $\{(a, b_2, t_1, t_2, h_3), (a, b_2, t_1, h_2, t_3)\} \subset W$ and on $\{(a, b_2, t_1, h_2, h_3), (a, b_2, t_1, t_2, t_3)\} \subset W$ is not determined by the first three entries a, b_2, t_1 only, but is also dependent on the last two.

Suppose we want to consider the objective chance that the second coin lands heads or tails. Then we set

$$\Sigma_{\text{chancy}} = \{ \emptyset, \ [H_2], \ [T_2], \ W \}$$

where H_2 (or T_2) denotes the proposition that the outcome of the second coin toss is heads (or tails), and where now $[H_2] = [H_2]_1 \cup [H_2]_2$ with

$$[H_2]_1 = \left\{ (x, y_1, y_2, y_3) \in W \mid x = b_1, \ y_2 = h_2 \right\}$$

and

$$[H_2]_2 = \left\{ (w, x, y_1, y_2, y_3) \in W \mid w = a, \ y_2 = h_2 \right\}$$

and similarly for $[T_2]$.

Suppose we are concerned about the objective chance of the second coin toss of the new coin. But at what time? It cannot of course be considered at step 0, because it is not known how it is decided whether the new coin shall be biassed or not. However, it could be considered at step 1, after the first decision was made, because the rest relies on coin tosses with objective chance. Or, another natural time is once the new coin is forged, and before it is used. However, on W_1 this happens at step 1, while on W_2 this toss happens at step 2. Or, we could also consider it at the time just before the coin is thrown for the second time.

3.3.2.1   After the First Step

In this case we set the time to $\tau := 1$. This means that it has been decided whether a fair or a biassed coin is to be forged, but nothing else. Clearly, the algebra representing the environment of the experiment is

$$\Sigma_{env} = \Sigma_1 = \{ \emptyset, \ W_1, \ W_2, \ W \}$$

As on W_2, the bias might be two kinds, each with probability $1/2$, the chance distribution is

$$ch_1(w, [H_2]) = \begin{cases} 0.5 \text{ if } w \in W_1 \\ 0.5 \cdot 0.35 + 0.5 \cdot 0.8 = 0.575 \text{ if } w \in W_2 \end{cases}$$

And $ch_1(w, [T_2]) = 1 - ch_1(w, [H_2])$.

Exactly the same way as we did in Sect. 3.3.1.4, here too we can extend chance onto the algebra generated by Σ_{env} and Σ_{chancy}, denoted by $\overline{\Sigma_{env} \cup \Sigma_{chancy}}$, but here this set is not the entire Σ.

Since after the first step all outcomes are assumed to be chancy, namely, the decision of the bias via a fair coin and then the coin toss three times, we can actually define objective chance for all those events. In this case we can consider the objective chance distribution on the event algebra:

$$\Sigma_{chancy} := \overline{\{\emptyset, [B_2], [B_3], [H_1], [T_1], [H_2], [T_2], [H_3], [T_3], W\}}$$

where

$$[B_2] := \left\{ (w, x, y_1, y_2, y_3) \,\middle|\, w = a, x = b_2, \ (y_i = t_i) \vee (y_i = h_i) \right\}$$

$$[B_3] := \left\{ (w, x, y_1, y_2, y_3) \,\middle|\, w = a, x = b_3, \ (y_i = t_i) \vee (y_i = h_i) \right\}$$

Note that $[B_1] = W \backslash ([B_2] \cup [B_3])$, so $[B_1] \in \Sigma_{\text{chancy}}$. In fact, Σ_{chancy} in this case is just the full Σ. Clearly, $ch_1(w, [B_1]) = 1$, $ch_2(w, [B_2]) = ch_1(w, [B_3]) = 0$ if $w \in W_1$, $ch_1(w, [B_1]) = 0$, $ch_2(w, [B_2]) = ch_1(w, [B_3]) = 0.5$ if $w \in W_2$, $ch_1(w, [B_1]) \cap [H_1] \cap [H_2] \cap [H_3]) = 0.5^3$ if $w \in W_1$, and so on.

3.3.2.2 At Some Stopping Time

It also make sense to consider objective chance after the coin is forged. However, this happens at different times on different possible worlds. So consider the following random variable:

$$\tau(w) = \begin{cases} 1 \text{ on } W_1 \\ 2 \text{ on } W_2 \end{cases}$$

Such a function is called *stopping time* in stochastic processes. Namely, given a filtration $(\Sigma_n)_{n \in \mathbb{N}}$ on a probability space W, a random variable $\tau : W \to \mathbb{N}$ is a stopping time, if for all $n \in \mathbb{N}$, the set where τ is less than or equal to n, $[\tau \leq n] = \{w \in W : \tau(w) \leq n\}$ is in Σ_n. Such a stopping time determines an event algebra:

$$\Sigma_\tau := \overline{\bigcup_{n \in \mathbb{N}} \left\{ E \mid E \in \Sigma_n \wedge E \subseteq [\tau \leq n] \right\}}$$

In our case,

$$\Sigma_\tau = \{ \emptyset \,, [B_1] \,, [B_2] \,, [B_3] \,, [B_1] \cup [B_2] \,, [B_2] \cup [B_3] \,, [B_1] \cup [B_3] \,, W \}$$

(that is, $W_1 = [B_1]$ is not split, but W_2 is split into $[B_2]$ and $[B_3]$). Let now $\Sigma_{\text{env}} := \Sigma_\tau$. Again, if we take Σ_{chancy} to be

$$\Sigma_{\text{chancy}} := \{ \emptyset \,, [H_2] \,, [T_2] \,, W \},$$

then the same way as before, on W_1:

$$ch_\tau((b_1, y_1, y_2, y_3), [H_2]) := 0.5$$

and

$$ch_\tau((a, x, y_1, y_2, y_3), [H_2]) := \begin{cases} 0.35 \text{ if } x = b_2 \\ 0.8 \text{ if } x = b_3 \end{cases}$$

It is easy to see that ch_τ is measurable with respect to $\Sigma_{\text{env}} = \Sigma_\tau$.

3.3.2.3   Observations on the Structure

In this section, we looked at a case where objective chance comes from a stochastic process (but in which not all branchings have associated probabilities). In case of a stochastic process, for each time t, there is an associated event algebra Σ_t representing the events that happened until time t. As more events happen until a later time, $\Sigma_{t_1} \subseteq \Sigma_{t_2}$ for $t_1 < t_2$. When we consider an event at a certain time determined by the past, that certain time might vary from possible world to possible world. Let τ be such function assigning to each possible world w a time $\tau(w)$. As τ is determined by the past, it is necessary that for each time value t, the set $[\tau \leq t] := \{w \in W | \tau(w) \leq t\}$ is in Σ_t. Such a τ is called stopping time. Σ_τ contains all events that happen until τ. In case of objective chance ch_τ (of chancy events Σ_{chancy}) determined by the past until τ, the environmental algebra Σ_{env} equals Σ_τ. Accordingly, for each $A \in \Sigma_{\text{chancy}}$, the function $ch_\tau(\cdot, A)$ must be measurable with respect to $\Sigma_{\text{env}} = \Sigma_\tau$. And of course, for each $w \in W$, the function $ch_\tau(w, \cdot)$ is a probability measure over Σ_{chancy}. Again, the notion of filtration and stopping time are missing from the literature on the Principal Principle and objective chance, but they are really essential for understanding time-dependent objective chance.

3.3.3   Is It Always the Past?

In the previous examples, the environmental algebra Σ_{env} always agreed with a Σ_τ algebra representing the events of the past with respect to a stopping time τ. This however does not have to be so. In fact, even Lewis (1994) later considered objective chance that depended on the future. What matters is how we set up our problem: what is given, and what is left chancy. It makes perfect sense to talk about the objective chance of a coin toss in the past, having fixed some conditions at a later time. It makes perfect sense to consider objective chance relative to a Σ_{env} other than Σ_τ for some stopping time τ. Consider again the example in Sect. 3.3.2.2. Having forged the coin, knowing whether it was forged biassed or not, and already seeing the result of the first toss, the agent knowing the process but not knowing which bias the coin has in his world, he can consider the objective chance of not only the second coin toss, but also the objective chance of having forged the first, second or the third type of coin. Why not? Consider the forging process described in Sect. 3.3.2.2. Given that the coin was forged biassed and the outcome of the first coin toss is heads, nothing prevents us to compute the objective chance of having a second type coin or a third type coin, and the objective chances of the outcomes of the next coin toss. Given that the coin was forged fair and that the outcome of the first toss is heads, we again have no problem telling the chance of having a third-type coin: it is 0.

Accordingly, we fix Σ_{env} the following way:

$$\Sigma_{\text{env}} := \overline{\{ \varnothing, [H_1^f], [T_1^f], [H_1^b], [T_1^b], W \}}$$

where the index stands for fair and biassed. $[H_1^f]$ is the event that the coin is fair and the first toss resulted heads, $[H_1^b]$ is the event that the coin is biassed and the first toss resulted heads, and similarly for $[T_1^f]$ and $[T_1^b]$. Clearly,

$$[H_1^f] := \left\{ (x, y_1, y_2, y_3) \in W \;\middle|\; x = b_1, \; y_1 = h_1 \right\}$$

and

$$[H_1^b] = \left\{ (w, x, y_1, y_2, y_3) \in W \;\middle|\; w = a, \; y_1 = h_1 \right\}$$

and similarly for $[T_1^f]$ and $[T_1^b]$. In other words, the outcome of the first coin toss as well as fairness and biassedness of the coin can be distinguished in Σ_{env}, but not whether the coin was forged to be second or third type as each element in Σ_{env} contains both versions or neither.

Since we consider the chance after the first coin toss,

$$\tau(w) = \begin{cases} 2 \text{ on } W_1 \\ 3 \text{ on } W_2 \end{cases}$$

and the atoms of Σ_τ are those sets of quadruples in W_1 or quintuples in W_2 for which the first two or three elements are fixed while the last two both range through heads and tails. Clearly,

$$\Sigma_{\text{env}} \neq \Sigma_\tau.$$

If we want to consider the chances of the various biasses as well as the chances of the outcomes of the second coin toss, we set

$$\Sigma_{\text{chancy}} := \overline{\{ \varnothing, [B_1], [B_2], [B_3], [H_2], [T_2], W \}}$$

In order to make the computations easier, let us modify the biasses for $[B_2]$ and $[B_3]$ from the previous examples. Suppose that the coin of the second kind has a chance to land heads by $0.4 = 2/5$, and the coin of the third kind has a chance to land heads by $0.75 = 3/4$. Otherwise the manufacturing process is the same as in Sect. 3.3.2.2.

For any $w \in W$ and $A \in \Sigma_{\text{chancy}}$, we have to compute $ch(w, A)$. As this in its first variable has to be measurable with respect to Σ_{env}, it is constant on the atoms of Σ_{env}, namely, on $[H_1^f]$, $[T_1^f]$, $[H_1^b]$, $[T_1^b]$. What is, for example $ch(w, [B_2])$ when $w \in [H_1^f]$? In this case w is a possible world on which the coin is fair. So it could not be manufactured the second way, and $ch(w, [B_2]) = 0$. Similarly $ch(w, [B_2]) = 0$ when $w \in [T_1^f]$. What is $ch(w, [B_2])$ when $w \in [H_1^b]$? Given the coin was manufactured to be biassed, the chance that it lands heads is $1/2 \cdot 2/5 + 1/2 \cdot 3/4 = 1/5 + 3/8 = 23/40$. Given the coin was manufactured to be biassed, the chance that it is manufactured in the second way and that it lands heads $1/2 \cdot 2/5 = 1/5$. So, $ch(w, [B_2]) = (1/5)/(23/40) = 8/23$ when $w \in [H_1^f]$. Similarly, we have $ch(w, [B_2]) = (1/2 \cdot 3/5)/(1/2 \cdot 3/5 + 1/2 \cdot 1/4) = (3/10)/(17/40) = 12/17$ when $w \in [T_1^b]$ So we have

$$ch(w, [B_2]) = \begin{cases} 0 \text{ if } w \in [H_1^f] \cup [T_1^f] = W_1 \\ 8/23 \text{ if } w \in [H_1^b] \\ 12/17 \text{ if } w \in [T_1^b] \end{cases}$$

Similarly the values for B_3 can be computed. Moreover,

$$ch(w, [H_2]) = \begin{cases} 0.5 \text{ if } w \in [H_1^f] \cup [T_1^f] = W_1 \\ \alpha \text{ if } w \in [H_1^b] \\ \beta \text{ if } w \in [T_1^b] \end{cases}$$

Where

$$\alpha = (1/2 \cdot 2/5 \cdot 2/5 + 1/2 \cdot 3/4 \cdot 3/4)/(2/5 \cdot 1/2 + 3/4 \cdot 1/2) \qquad (3.1)$$

and

$$\beta = (1/2 \cdot 3/5 \cdot 3/5 + 1/2 \cdot 1/4 \cdot 1/4)/(3/5 \cdot 1/2 + 1/4 \cdot 1/2)$$

Similarly for T_2.

Consequently, it is easy to construct examples when objective chance relative to some event algebra Σ_{env} is considered with Σ_{env} different from Σ_τ for any stopping time τ. The notion of time may in fact be entirely missing from the model, objective chance can still be made sense. So for the objective chance it is not Σ_τ that is important, but Σ_{env} expressing the events relative to which objective chance is considered. Σ_{env} may or may not agree with some Σ_τ.

To make this point even stronger, consider the situation when we can allow various space-time models. In other words, some of the possible worlds correspond to certain space-time models, others correspond to other space-models. Then we can ask, given a model, what is the objective chance that something happens in that model. Clearly, this is not a chance associated to a certain time, but just to various space-time models. In such a case Σ_{env} would contain sets of space-time models.

3.3.4 Impossible Events Have Zero Chance

There is a very important property of objective chances that we have not talked about yet, and that is entirely missed by both Lewis's original paper as well as by later literature on the Principal Principle. Namely, consider an $E \in \Sigma_{\text{env}}$ and an $A \in \Sigma_{\text{chancy}}$ such that $E \cap A = \emptyset$. In other words, if we fix the chancy setup to satisfy the properties of E, then there is no way for event A to come about because none of the possible worlds in E are also in A. Therefore an objective chance function must not give non-zero chance of A for possible worlds in E. But assuming only that ch is measurable with respect to Σ_{env} in its first variable and that it is a probability distribution over Σ_{chancy} in its second variable, ch might still violate this condition. So we have to additionally require it for an objective chance. Consider the following specific example.

Let us go back now to the original example, but suppose we add a single further possible world to it:

$$W := \left\{ (x, y) \;\middle|\; (x = b_1 \vee x = b_2 \vee x = b_3) \wedge (y = h \vee y = t) \right\} \cup \{(b_4, h)\}$$

In other words, with b_4 bias, the coin only lands on heads. For example, both sides of the coin is heads. With this modification,

$$\Sigma_{\text{env}} = \overline{\{\emptyset, [B_1], [B_2], [B_3], [B_4]\}}$$

with $[B_4] = \{(b_4, h)\}$, and

$$[T] = \{(b_1, t), (b_2, t), (b_3, t)\}, \qquad [H] = \{(b_1, h), (b_2, h), (b_3, h), (b_4, h)\}$$

Clearly, $[B_4] \cap [T] = \emptyset$. That is no surprise, because in our setup, if the coin has the b_4 bias, it cannot land heads. So is it possible that $ch((b_4, h), [T]) \neq 0$? No it is not. The atom $[B_4]$ of Σ_{env} contains no elements in $[T]$, meaning that on $[B_4]$, the outcome of the chancy event can never be in $[T]$. Since this is an impossible event in $[B_4]$, it has no chance to occur in $[B_4]$, and hence the chance of $[T]$ in $[B_4]$ must be 0.

This can of course be told in general: if $E \in \Sigma_{\text{env}}$ and $A \in \Sigma_{\text{chancy}}$ is such that $E \cap A = \emptyset$, then the objective chance $ch(\cdot, A)$ must be 0 on E as there is no way A will occur on E. This is an additional requirement on objective chance.

3.3.5 Frequentist's Objective Chance

In this final example, we consider the frequentist's model of objective chance, and argue that it can also be described with two subalgebras, Σ_{env} and Σ_{chancy} as the other cases. Consider the situation when the set of possible worlds contains only

8 possible worlds: 3 independent tosses of a fixed coin, each with two possible outcomes, heads and tails. This set of possible worlds has the following elements:

$$W = \{(h,h,h),(h,h,t),(h,t,h),(t,h,h),(t,t,h),(t,h,t),(h,t,t),(t,t,t)\}$$

The event that the i'th coin toss is heads is

$$[H_i] = \{(y_1,y_2,y_3) \in W | \pi_i((y_1,y_2,y_3)) = h\}$$

that it is tails is

$$[T_i] = \{(y_1,y_2,y_3) \in W | \pi_i((y_1,y_2,y_3)) = t\}$$

We can take $\Sigma_{\text{chancy}} := 2^W$, as all events are chancy.

Now, a frequentist would say that the chance of heads, that is, of any of $[H_i]$ at time 0 is 1 on the possible world (h,h,h), is $2/3$ on (h,h,t), (h,t,h), and (t,h,h), it is $1/3$ on (t,h,h), (t,t,h), (t,h,t), and 0 on (t,t,t). Accordingly, since in this model the chance is determined by the frequencies of h and t on the possible worlds, Σ_{env} has to be set to

$$\Sigma_{\text{env}} := \overline{\{\{(h,h,h)\}, \{(h,h,t),(h,t,h),(t,h,h)\}, \{(t,t,h),(t,h,t),(h,t,t)\}, \{(t,t,t)\}\}}$$

and $ch_0((h,h,h),[H_i]) = 1$, $ch_0(w,[H_i]) = 2/3$ if $w \in \{(h,h,t),(h,t,h),(t,h,h)\}$, $ch_0(w,[H_i]) = 1/3$ if $w \in \{(t,t,h),(t,h,t),(h,t,t)\}$, $ch_0((t,t,t),[H_i]) = 0$. And clearly, $ch_0(w,[T_i]) = 1 - ch_0(w,[H_i])$. Now, as we set $\Sigma_{\text{chancy}} = 2^W$, we have to define $ch_0(w,S)$ for any subset of W. What is for example, $ch_0(w,[H_1] \cap [H_2])$? The first idea would be $ch_0(w,[H_1] \cap [H_2]) = ch_0(w,[H_1]) \cdot ch_0(w,[H_2])$ as the tosses are independent. However, we immediately realize, that as in this model, we assumed that the coin is tossed only 3 times, the tosses are not independent any more once we fix the chance.[3] For example, if the chance of heads is assumed to be $2/3$, and the first two tosses were both heads, then we know the third must be tails. The objection could be that before the third toss, we do not know that there will be no more tosses. However, that means that our model of the world is wrong, we have to allow possible worlds with any long sequences of tosses, and in that case, the longer the sequence is, the closer the chance will get to result independence between the 2nd and the 3rd coin toss. But given the frequentist view, they will only be perfectly independent if we assume infinitely long sequences.

So what is $ch_0(w,[H_1] \cap [H_2])$? Clearly, it is $ch_0(w,[H_1] \cap [H_2]) = ch_0(w,[H_2]|[H_1]) \cdot ch_0(w,[H_1])$. But what is $ch_0(w,[H_2]|[H_1])$? This can again be obtained by the frequentist view. For example, if $w \in \{(h,h,t),(h,t,h),(t,h,h)\}$, then given that the result of the first toss is h, the rest of the tosses are either ht or

---

[3]Hence we strongly disagree with the treatments of some other authors such as Pettigrew (1986) on page 6.

th, and hence $ch_0(w, [H_2] | [H_1]) = 1/2$. then for this w, $ch_0(w, [H_1] \cap [H_2]) = 1/3$, and not $4/9$. Similarly, $ch_0(w, [H_1] \cap [H_2] \cap [H_3]) = 0$, and so on.

It is easy to see that this frequentist objective chance does not correspond to any Σ_τ either, because chance depends on the whole history, but Σ_{env} is smaller than the full $\Sigma_4 = \mathcal{P}(W)$.

3.4   The Subjectivist Agent's Model of Objective Chance

According to the forgoing, I propose the following definition of a model of possible worlds with objective chance:

Definition 4.1 (Possible World Model with Objective Chance) $(W, \Sigma, \Sigma_{env}, \Sigma_{chancy}, ch)$ is a possible world model with objective chance if

- W is a set (of possible worlds)
- Σ is a σ-algebra on W
- Σ_{env} is a σ-subalebra of Σ
- Σ_{chancy} is a σ-subalebra of Σ
- $ch : W \times \Sigma_{chancy} \to [0, 1]$ such that

    - For all $w \in W$, $A \mapsto ch(w, A)$ is a probability measure on Σ_{chancy}
    - For all $A \in \Sigma_{chancy}$, $w \mapsto ch(w, A)$ is a Σ_{env}-measurable function on W
    - If for some $E \in \Sigma_{env}$ and $A \in \Sigma_{chancy}$, $E \cap A = \emptyset$, then $ch(w, A) = 0$ for all $w \in E$.

Of course, if for each time t we have some Σ_{env} and some ch, then we can index ch with time, but time is not essential for the definition of objective chance.

This definition works for infinite W as well. A sigma algebra Σ is closed under countably infinite union and complementation, and contains W.

Again, as in Sect. 3.3.1.4, for each w, the chance of any event in $\overline{\Sigma_{env} \cup \Sigma_{chancy}}$ can be defined with respect to Σ_{env}. Given $(W, \Sigma, \Sigma_{env}, \Sigma_{chancy}, ch)$, for each $w \in W$, there is a *unique* extension of $ch(w, \cdot)$ to $\overline{\Sigma_{env} \cup \Sigma_{chancy}}$ such that for the extended chance, which we shall denote by $\overline{ch}$ also satisfies the above requirements, and $(W, \Sigma, \Sigma_{env}, \overline{\Sigma_{env} \cup \Sigma_{chancy}}, \overline{ch})$ is a possible world model with objective chance: For all $w \in W$, $E \in \Sigma_{env}$, set

$$\overline{ch}(w, E) := \begin{cases} 1 \text{ if } w \in E \\ 0 \text{ if } w \notin E \end{cases}$$

Because of the last requirement in the definition of a possible world model with objective chance, there is no other way to define $\overline{ch}$ on Σ_{env}. If $w \notin E \in \Sigma_{env}$, then it is in $E^\perp \in \Sigma_{env}$, and since $E^\perp \cap E = \emptyset$, by the last requirement $\overline{ch}(w, E) = 0$ for all $w \in E^\perp$ as long as $\overline{ch}$ is an objective chance function. Then for any $A \in \Sigma_{chancy}$, $E \in \Sigma_{env}$, and $w \in E$,

$$\overline{ch}(w, E \cap A) = \overline{ch}(w, E) \cdot \overline{ch}(w, A|E)$$

must hold if $\overline{ch}(w,\cdot)$ is to be a probability distribution. As $w \in E$, $\overline{ch}(w,E) = 1$, and $\overline{ch}(w,A|E) = \overline{ch}(w,A)$. But as $\overline{ch}$ is an extension of ch, $\overline{ch}(w,A) = ch(w,A)$. Furthermore, when $w \notin E$, $\overline{ch}(w,E \cap A) = 0$. Hence I have shown that there is only one way of defining $\overline{ch}$, namely with the formula

$$\overline{ch}(w,E \cap A) = \overline{ch}(w,E) \cdot ch(w,A)$$

As all elements in $\overline{\Sigma_{\mathrm{env}} \cup \Sigma_{\mathrm{chancy}}}$ can be written as the union of sets of the form $E \cap A$ as above (the event algebra Σ is generated by Σ_{env} and Σ_{chancy}), this defines a unique probability measure $S \mapsto \overline{ch}(w,S)$ on $\overline{\Sigma_{\mathrm{env}} \cup \Sigma_{\mathrm{chancy}}}$.

The above rigorous definition of objective chance opens the possibility of investigating the Principal Principle and admissible evidence in a mathematically precise way, which I shall do elsewhere.

Acknowledgements I would like to thank to the organizers of the Budapest-Krakow Workshop on Probability, Causality and Determinism for granting me the opportunity to present my ideas about objective chance and the Principal Principle. I would also like to thank Leszek Wroński, Miklós Rédei, Zalán Gyenis, and Balázs Gyenis for the inspiring discussions and their insightful comments.

References

Bana, G. 2016. On the formal consistency of the principal principle. *Philosophy of Science* 83(5): 988–1001. University of Chicago Press.

Lewis, D. 1980. A subjectivist's guide to objective chance. In *Studies in inductive logic and probability*, ed. R.C. Jeffrey, vol. II, 263–293. Berkely: University of California Press.

Lewis, D. 1994. Humean supervenience debugged. *Mind* 103(412): 473–490. Oxford University Press.

Pettigrew, R. 2013. What chance-credence norms should not be. *Noûs* 73(3): 177–196. Wiley.

Rédei, M., and Z. Gyenis. 2016. Measure theoretic analysis of consistency of the principal principle. *Philosophy of Science* 83(5): 972–987. University of Chicago Press.

Chapter 4
Four Approaches to the Reference Class Problem

Christian Wallmann and Jon Williamson

Abstract We present and analyse four approaches to the reference class problem. First, we present a new objective Bayesian solution to the reference class problem. Second, we review Pollock's combinatorial approach to the reference class problem. Third, we discuss a machine learning approach that is based on considering reference classes of individuals that are similar to the individual of interest. Fourth, we show how evidence of mechanisms, when combined with the objective Bayesian approach, can help to solve the reference class problem. We argue that this last approach is the most promising, and we note some positive aspects of the similarity approach.

4.1 Introduction

The problem of determining the probability that a particular patient has a certain attribute (e.g., has a certain disease, or has a certain prospect of recovery) is of fundamental importance to medical diagnosis, prediction and treatment decisions. Theories of *direct inference* aim to solve this problem of the single-case. In direct inference, single-case probabilities are often calibrated to estimates of chances in reference classes to which the individual of interest belongs. The major problem in direct inference is to determine an appropriate single-case probability when an individual belongs to several reference classes for which data is available, and where estimates of chances differ from reference class to reference class. This is the *reference class problem*.

Let us consider how direct inference sometimes proceeds by means of an example. The question is whether Nataly, a patient with breast cancer, will survive at least five more years. The doctor may know the size of the index lesion (S), node status (N) and (G) grade of tumour of the patient. She then calculates the Nottingham prognostic index (NPI) score by $NPI = [0.2 \times S] + N + G$ (Haybittle et al. 1982). Let's suppose Nataly has an NPI-score of 4.2. There is statistical

C. Wallmann (✉) • J. Williamson
Department of Philosophy, University of Kent, Canterbury, UK
e-mail: c.wallmann-520@kent.ac.uk

© Springer International Publishing AG 2017   61
G. Hofer-Szabó, L. Wroński (eds.), *Making it Formally Explicit*, European Studies
in Philosophy of Science 6, DOI 10.1007/978-3-319-55486-0_4

evidence that 7 out of 10 people with an NPI-score between 3.4 and 5.4 survive for more than 5 years. One might infer, then, that Nataly has a probability of survival of 70%.

As straightforward as this may seem, in practice it is rather difficult to determine appropriate reference classes. Often an individual belongs to many populations for which there is statistical information about the chance of the attribute of interest in those populations. In our example, the patient will have a certain nationality, a certain attitude towards religion, a certain social status, and a certain genetic profile, and there may be evidence of chances available for several of these reference classes. While we might assume that religious belief is irrelevant for survival time, this is less clear for nationality and social status, and certainly false for genetic profile. The most intractable problem for direct inference is the problem of conflicting reference classes. The problem occurs when chances are available for two (or more) reference classes to which the individual of interest belongs, these chances differ, and there is no chance available for the intersection of these reference classes. John Venn remarked with respect to John Smith, an Englishman that has tuberculosis,

> Let us assume, for example, that nine out of ten Englishmen are injured by residence in Madeira, but that nine out of ten consumptive persons are benefited by such a residence. These statistics, though fanciful, are conceivable and perfectly compatible. John Smith is a consumptive Englishman; are we to recommend a visit to Madeira in his case or not? In other words, what inferences are we to draw about the probability of his death? Both of the statistical tables apply to his case, but they would lead us directly contradictory conclusions. [...] Without further data, therefore we can come to no decision. (Venn 1888, p. 222–223)

Suppose that we know that an individual c belongs to two reference classes B and C, written Bc and Cc. Suppose further that we know the chance of the target attribute A in the reference class B as well as in the reference class C, i.e., $P^*(A|B) = r$ and $P^*(A|C) = s$ (and nothing else). The problem of conflicting reference classes is the problem of determining the probability $P(Ac)$ that c has attribute A.

A word on notation. We use P^* to represent the objective chance distribution. This is *generic* in the sense that it is defined over attributes, classes and variables which are repeatedly instantiatable. We make no metaphysical assumptions about chance here: a chance might be understood as a dispositional attribute (or propensity), a long-run frequency, an objectivised subjective probability, or posited by a Humean account of laws, for example. We use *freq* to denote the sample frequency. Again, this is generic. It is the probability distribution induced by a sample or a dataset, and it may be used to estimate the chance distribution P^*, i.e., the data-generating distribution. Finally, we take other probability functions, such as P, $P^†$, to be *single-case* probability functions, to be used for direct inference. These are single-case in the sense that they are defined over propositions or events which are not repeatedly instantiable.

In Sect. 4.2, we develop a new objective Bayesian solution to the problem of conflicting reference classes. In Sect. 4.3, we review Pollock's approach to the problem. We show that it is based on mistaken assumptions. In Sect. 4.4, we relate similarity-based machine learning techniques to the reference class problem. All these approaches are classifiable as generic-probability approaches: chances in reference classes are estimated by frequencies induced by datasets and those estimates are then aggregated to obtain a probability for the single case. In Sect. 4.5, we briefly discuss two key challenges that face generic-probability approaches: the problem of small sample size, which arises when reference classes are so narrowly defined that it is difficult to obtain samples to estimate the corresponding chances, and the problem of inconsistent marginals, which arises when different samples yield incompatible sample frequencies. We show that the objective Bayesian approach can address these challenges by appealing to confidence region methods, and that the similarity approach can be re-interpreted as a single-case-probability approach in order to avoid these difficulties. In Sect. 4.6, we show how to make use of evidence of mechanisms to help solve the reference class problem. Evidence of mechanisms helps in two ways: it can help by enriching the structure of the problem formulation, and it can help to extrapolate evidence of chances from reference classes that are not instantiated by the particular individual of interest to those that are. Either way leads to more credible direct inferences. We conclude by discussing the circumstances in which the various methods are appropriate in Sect. 4.7.

4.2 An Objective Bayesian Approach

According to the version of objective Bayesian epistemology developed by Williamson (2010), one can interpret predictive probabilities as rational degrees of belief, and these rational degrees of belief are obtained by taking a probability function, from all those that satisfy constraints imposed by evidence, that has maximal entropy.[1] That degrees of belief should be probabilities is called the Probability norm. That they should satisfy constraints imposed by evidence is the Calibration norm. In particular, degrees of belief should be calibrated to chances, insofar as the evidence determines these probabilities. That a belief function should have maximal entropy is the Equivocation norm. The maximum entropy function is interpretable as the most equivocal or least committal probability function (Jaynes 1957). The Equivocation norm is justifiable on the grounds of caution: a maximum entropy function is a probability function which minimises worst-case expected loss (Williamson 2017a, Chapter 9).

Williamson (2013) locates the reference class problem at the stage of the Calibration norm:

> The infamous reference class problem must be tackled at this stage, i.e., one must decide which items of evidence about the generic physical probabilities should be considered when determining single case probabilities [...]. (Williamson 2013, p. 299)

---

[1]The entropy of a probability function P defined on a set of logically independent propositions $\{E_1, \ldots, E_n\}$ is defined by $-\sum_{i=1}^{n} P(E_i) \log P(E_i)$.

To solve the problem of conflicting reference classes, Williamson draws on other approaches to direct inference. Without endorsing it, he discusses Kyburg's approach as one possible option. A combination of Williamson's and Kyburg's approach first combines the information in conflicting reference classes to obtain an interval for $P(Ac)$ and then applies the Equivocation norm to this already aggregated degree of belief. $\{P^*(A|B) = 0.9, P^*(A|C) = 0.4, Bc, Cc\}$, for instance, yields by Calibration $P(Ac) \in [0.4, 0.9]$ and subsequently by Equivocation $P(Ac) = 0.5$. However, as we are going to see now, the objective Bayesian approach has sufficient resources to solve the problem of conflicting reference classes on its own. Rather than considering it purely as a calibration problem, the proposal presented here spreads the load between the Calibration norm and the Equivocation norm.

The objective Bayesian approach presented here proceeds first by calibrating, then by aggregating. First, it calibrates conditional degrees of belief to estimates of chances, $P(Ac|Bc) = P^*(A|B) = r$ and $P(Ac|Cc) = P^*(A|C) = s$. Second, it equates the direct inference probability that c belongs to the target class A with $P^\dagger(Ac|Bc \wedge Cc)$, where $P^\dagger$ satisfies the constraints $P^\dagger(Ac|Bc) = r$ and $P^\dagger(Ac|Cc) = s$ but is otherwise as equivocal as possible. We then have

New objective Bayesian solution to the problem of conflicting reference classes. $P(Ac) = P^\dagger(Ac|Bc \wedge Cc)$, where $P^\dagger$ is a probability function that has maximal entropy among all probability functions P that satisfy $P(Ac|Bc) = r$ and $P(Ac|Cc) = s$.

The objective Bayesian approach combines the maximum entropy principle with probabilistic logic. The problem of determining the direct inference probability can be solved by linear programming and optimization techniques. If $P^*(A|B) = r$ and $P^*(A|C) = s$, the solution to the reference class problem is given by $P^\dagger(Ac|Bc \wedge Cc) = \frac{x_1}{x_1 + x_5}$ where the vector x is the solution to the following optimization problem[2]:

$$\text{Maximise} - \sum_{i=1}^{8} x_i \log x_i \text{ subject to } Sx = b \text{ and } x_i \geq 0, \text{ where } b = \begin{pmatrix} 0 \\ 0 \\ 1 \end{pmatrix} \text{ and } S =$$

$$\begin{pmatrix} r-1 & r-1 & 0 & 0 & r & r & 0 & 0 \\ s-1 & 0 & s-1 & 0 & s & 0 & s & 0 \\ 1 & 1 & 1 & 1 & 1 & 1 & 1 & 1 \end{pmatrix}.$$

Although it can be difficult to provide analytic solutions to such problems, they can be solved numerically, by using, for instance, MAPLE or Matlab software.

To give an example, $\{P^*(A|B) = 0.9, P^*(A|C) = 0.4, Bc, Cc\}$ leads to $P(Ac) = 0.83$. Note that the objective Bayesian approach does not necessarily assign a very equivocal degree of belief to the proposition Ac. Indeed, it can lead to degrees of belief that are more extreme than either of the reference class frequencies. For instance, in absence of further constraints, $P^*(A|B) = 0.9, P^*(A|C) = 0.9$ leads to $P(Ac) = 0.96$. The reason for this is that the objective Bayesian approach leads

---

[2]See Wallmann and Kleiter (2014a,b) for a general procedure for generating the relevant optimization problem.

to beliefs that are equivocal *on average*. It may assign extreme degrees of belief to certain propositions and instead assign less extreme degrees of belief to other propositions in order to maximise the extent to which the belief function as a whole is equivocal.

In the next section, we will consider an approach that claims that more can be done to constrain $P^*(A|B \wedge C)$ before calibrating.

4.3 Pollock's Approach

Pollock's approach to direct inference involves first aggregating, then calibrating. Pollock first aggregates the values of the conflicting reference classes and estimates the value of $P^*(A|B \wedge C)$, and then calibrates $P(Ac)$ to the result. Since in Pollock's theory $P^*(A|B \wedge C)$ can be very well estimated (with probability 1), there is no role for equivocation to play.

Pollock motivates his theory of direct inference this way:

> Suppose we have a set of 10,000,000 objects. I announce that I am going to select a subset, and ask you how many members it will have. Most people will protest that there is no way to answer this question. It could have any number of members from 0 to 10,000,000. However, if you answer, "Approximately 5,000,000", you will almost certainly be right. This is because, although there are subsets of all sizes from 0 to 10,000,000, there are many more subsets whose sizes are approximately 5,000,000 than there are of any other size. In fact, 99% of the subsets have cardinalities differing from 5,000,000 by less than .08%. (Pollock 2011, p. 329)

This "peaking" property holds for finite sets and follows from elementary combinatorics. Moreover, the distribution of the subsets gets needle-like in the limit: the larger the set is, the greater the proportion of subsets that have size close to half the size of the set. Pollock takes this fact as a starting point for his theory of nomic probability. Pollock calls his theory 'nomic' because he is concerned with probabilities that are involved in statistical laws. Rather than being concerned with frequencies and relative frequencies involving actual events he is concerned with frequencies and relative frequencies among physically possible worlds. Probabilities therefore contain an irreducible modal element.

As a natural generalization from finite sets to infinite sets, Pollock's theory of nomic probabilities assumes that these peaking properties hold for infinite sets with probability 1.[3] If a peaking probability holds with probability 1, then the value

---

[3]Let pe be a point. It is called peaking point with probability 1 iff for all $\delta > 0$, $PROB(|P^*(A|S) - pe| < \delta) = 1$, i.e., for all $\epsilon > 0$, $PROB(|P^*(A|S) - pe| < \delta) > 1 - \epsilon$. If we think of δ being very small, for instance, 0.000001, then this means that almost all subsets S are such that $pe - 0.000001 \leq P^*(A|S) \leq pe + 0.000001$. Note that probability 1 does not mean that there are no exceptions, i.e., even if $PROB(|P^*(A|S) - pe| < \delta) = 1$, there are S such that $|P^*(A|S) - pe| > \delta$. However, such S are comparably few.

around which the nomic probabilities peak is called the *expectable value*. This leads
to the following default independence principle:

If $P^*(A|B) = x$, then the expectable value for $P^*(A|B \wedge C)$ is x.

Pollock's solution to the problem of conflicting reference classes is to set
the direct inference probability to the expectable value of the target attribute in
the intersection of the conflicting reference classes (Pollock 2011). Moreover, he
calculates the relevant expectable value. Let $P^*(A|B)$ denote the nomic probability
of A given B (Pollock 2011). If, in addition to the reference class probabilities
$P^*(A|B) = r$ and $P^*(A|C) = s$, the "base rate" $P^*(A|U) = a$ is given where
$B, C \subseteq U$, Pollock shows that the expectable value of $P^*(A|B \wedge C)$ exists and is
given by

$$Y(r, s|a) = \frac{rs(1 - a)}{a(1 - r - s) + rs} \ .$$

The Y-function can be used to tackle the problem of conflicting reference classes.
If there is no knowledge of the chance of the target attribute A in some joint upper
class U of B and C—i.e., if for no such U $P^*(A|U)$ is available—its expectable value
p_0 can still be determined (Pollock 2011). Since the expectable value is attained with
probability 1, the degree of under-determination for $P^*(A|B \wedge C)$ is very small and
ignoring the Equivocation desideratum seems to be reasonable, i.e., we may equate
$P(Ac)$ with the expectable value for $P^*(A|B \wedge C)$.

Pollock's solution to the problem of conflicting reference classes. The direct inference
probability is given by

$$P(Ac) = Y(r, s|p_0) \ .$$

The expectable value p_0 of $P^*(A|U)$ is given by the first component of the
solution (a, b, c) to the following system of equations (Pollock 2011).

$$\left(\frac{1 - r}{1 + (r - p_0)b}\right)^{1-r} \left(\frac{r}{-br + p_0}\right)^r = 1$$

$$\left(\frac{1 - s}{1 + (s - p_0)c}\right)^{1-s} \left(\frac{s}{-cs + p_0}\right)^s = 1$$

$$2p_0^3 - (-2br - 2cs + b + c - 3)p_0^2$$
$$+ (2bcrs - bcr - bcs + bc + 2br + 2cs - b - c + 1)p_0 - cbrs = 0$$

Although it is often difficult to provide explicit formulae for expectable values, they
can be calculated numerically, by using Pollock's LISP-code (Pollock 2016).

Pollock's expectable value is relative to the probability distribution that is used
to calculate the expectable value. The quality of the expectable value depends
on the accuracy of this probability distribution. For his probability distribution,

Pollock extrapolates simple combinatorial probabilities from the finite to the infinite case. Doubt has been raised whether combinatorial probabilities are accurate in the domain of reference class reasoning (Wallmann 2017). Especially, the fact that $P^*(A|B \wedge C)$ is almost certainly very close to $Y(r, s|p_0)$ is difficult to reconcile with experience. In practice, we often find subsets that have a rather different chance of the target attribute than the original set. For instance, smoking rates in the United States vary strongly with gender, age, education, poverty status and many more. But according to Pollock's nomic probabilities, such variations seem to be almost impossible.

The mistake is this: the combinatorial probabilities in the finite case consider *arbitrary* subsets of the sets B and C. Every subset has the same relevance for direct inference. However, in the context of direct inference this is unreasonable. If we use a certain subset of B in practice, most likely we will not use a different but almost identical subset for direct inference. For instance, if we use the set of all Austrians, most likely we will not use the set of all Austrians except for Alexander for direct inference. Being Austrian but being not identical with Alexander is not expressing any causally relevant attribute. Thus, not every subset of a reference class is itself a reference class. We tend to consider classes of individuals that instantiate natural attributes—attributes which are causally relevant to the attribute of interest. Now, causally relevant attributes tend to be difference makers, i.e., they tend to be probabilistically dependent. Therefore, real reference-class probabilities vary to a greater extent than arbitrary-subset probabilities. Instead of clustering very closely around the expectable value, frequencies within sub-reference classes are more likely to cluster around multiple different values. Although expected values exist in the case of sub-reference classes, expectable values do not. While peaking properties hold with probability 1 for arbitrary subclasses, this fact is irrelevant to direct inference. To take another example, the proportion of smokers varies to a great extent between sub-reference classes of the reference class of all people living in the United States. The fact that most subsets of all people living in the United States share the same proportion of smokers with all people living in the United States is irrelevant to direct inference, because in direct inference we are only concerned with natural attributes as, for instance, gender, age, educational level. Peaking properties do not hold with probability 1 for classes that correspond to such natural attributes.

4.4   The Similarity Approach

Suppose that a reasonable measure of similarity between two individuals is available. The basic idea behind the similarity approach to direct inference is simple: we may predict whether an individual c will get a certain disease by considering the chance of disease in reference classes of individuals similar to the individual c. The more similar a reference class of individuals is to c, the more relevant information about the chance of disease in this class is for predicting whether c will get the disease.

An attribute-based similarity measure is based on the number of shared attributes. One way to define a attribute-based similarity measure is Gower's similarity coefficient (Gower 1971).[4] This measures the number of shared attributes of individuals in a reference class R and an individual c. Suppose that $R_1, \ldots, R_n$ are reference classes. Let $C_1, \ldots, C_N$ be the attributes that c is known to have (the C_i's may also contain negations). Let $C_j(R)$ denote the fact that all individuals in the reference class R have the attribute C_j.

$$sim_{Gow}(R, c) = \frac{|\{j \in \{1, \ldots, N\} : C_j(R)\}|}{N} \tag{4.1}$$

The attribute similarity solution to the reference class problem is given by:

Attribute similarity solution to the reference class problem. For $i = 1, \ldots, n$, let $P^*(A|R_i) = x_i$. Then the direct inference probability that c has disease A is given by

$$P(Ac) = T \sum_{i=1}^{n} sim_{Gow}(R_i, c)x_i, \tag{4.2}$$

where T is a normalizing constant. For instance, if Bc, Cc, Hc, Ec and $P^*(D|\neg E \wedge C) = x, P^*(D|B \wedge \neg C \wedge E \wedge H) = y, P^*(D|B \wedge C) = z$, then $P(Dc) = \frac{4}{6}(\frac{1}{4}x + \frac{3}{4}y + \frac{2}{4}z)$.

Observe that the direct inference probability may be influenced by chances in reference classes to which the individual does not belong. Hence, the similarity approach aims to solve a problem even more general than the problem of conflicting reference classes—the problem of applying chances in arbitrary classes to specific individuals. We shall return to this problem when we discuss extrapolation in Sect. 4.6.2.

Attribute-based similarity measures are commonly used in machine learning. Indeed, the approach advocated here is a special case of the machine learning technique called k-nearest neighbour weighted mean imputation; for details see Jerez et al. (2010, pp. 110–111). To impute a missing data value, a weighted average value of the k most similar reference classes is taken. Here, $k = n$, i.e., all reference classes are similar enough to contribute to the weighted average. More sophisticated similarity measures incorporate, for instance, base rates of diseases in the general population (Davis et al. 2010). However, attribute-based similarity measures do not distinguish between causally relevant and irrelevant attributes; every attribute is equally important.

---

[4]For a related but more sophisticated similarity measure see Davis et al. (2010).

4.5 Generic-Probability vs Single-Case-Probability Approaches

The objective Bayesian and Pollock's approach are generic-probability approaches. They follow the following procedure:

1. Group individuals to classes (reference classes).
2. Estimate the chance of the attribute of interest in each class (a generic probability).
3. Determine a direct inference probability from these values by maximum entropy or other techniques.

Generic-probability approaches face two fundamental challenges. One key obstacle is what we call the *problem of small sample size*. On the one hand, we would prefer *narrow* reference classes, i.e., classes of individuals which share many of the features instantiated by the individual in question, because such classes are particularly relevant to the individual. On the other hand, however, a narrow reference class is likely to contain few sampled individuals. Where this sample size is small, the sample frequency is likely to be rather different from the true chance that it is supposed to estimate. Thus a narrow reference class tends to yield an inaccurate estimate of the chance of the attribute of interest within the reference class. More generally, to arrive at a precise estimate of the data-generating chance distribution defined over many variables, a very large number of observations is needed, because a relatively small proportion of sampled individuals will share any particular combination of values of measured variables. A dataset measuring a large number of variables will often be too small to provide a reasonable estimate of the data-generating chance function.

A second key obstacle is what we call the problem of *inconsistent marginal probabilities*. This occurs when several samples are collected—several datasets are obtained—and certain variables occur with different frequencies in different datasets: there is then no joint probability function whose marginal probabilities match all the distributions determined by the datasets. This problem is very common. It may be attributable to bias, chance or to small sample sizes.

The following example, pitched at the objective Bayesian solution to the reference class problem, illustrates the two challenges.

- Suppose datasets D_1, D_2, D_3 measure sets of variables $V_1 = \{X_1, X_2, X_3\}, V_2 = \{X_2, X_3, X_4\}, V_3 = \{X_1, X_2, X_4\}$ respectively.
- The objective Bayesian approach seeks to find an appropriate joint probability function $P(X_1, X_2, X_3, X_4)$, by the following procedure:

  1. Marginals of the data-generating chance distribution P^* are estimated by the sample distributions determined by the datasets:

     – $P^*(X_1, X_2, X_3)$ is estimated by the observed frequency $freq_1(X_1, X_2, X_3)$ of D_1.

- $P^*(X_2, X_3, X_4)$ is estimated by the observed frequency $freq_2(X_2, X_3, X_4)$ of D_2.
- $P^*(X_1, X_2, X_4)$ is estimated by the observed frequency $freq_3(X_1, X_2, X_4)$ of D_3.

2. Consider the set $\mathbb{E}$ of all probability functions P that satisfy $P(X_1, X_2, X_3) = P^*(X_1, X_2, X_3)$, $P(X_2, X_3, X_4) = P^*(X_2, X_3, X_4)$, and $P(X_1, X_2, X_4) = P^*(X_1, X_2, X_4)$.
3. Determine the probability distribution $P^\dagger$ in $\mathbb{E}$ with maximum entropy.

The problem of small sample size arises in Step 1 if the datasets have too few observations to yield plausible estimates of the chances. The problem of conflicting marginals arises in Step 2. The dataset distributions $freq_1(X_1, X_2, X_3)$, $freq_3(X_1, X_2, X_4)$ determine the respective marginal distributions $freq_1(X_1, X_2)$, $freq_3(X_1, X_2)$. Typically, $freq_1(X_1, X_2) \neq freq_3(X_1, X_2)$, i.e., we have inconsistent marginals. Consequently, there is no probability function P that satisfies the above constraints.

The objective Bayesian approach has a potential line of response to these two challenges—a response that involves an appeal to *confidence regions* (Williamson 2017b, §4). As discussed above, the frequency distribution $freq_i$ determined by dataset D_i can be thought of as a point estimate of the marginal data-generating chance distribution $P^*(V_i)$, defined over the set V_i of variables measured by that dataset. For instance, $freq_1(X_1, X_2, X_3)$, the observed frequency distribution of dataset D_1, is treated as a point estimate of the data-generating chance distribution $P^*(X_1, X_2, X_3)$ in the above example. Now, a point estimate is almost always wrong. Rather than use a point estimate, one can instead infer that the data-generating chance distribution lies in a region around the point estimate—the confidence region R_i. This confidence region depends on the confidence level, i.e., how probable it is that similar samples will yield a sample distribution such that the chance distribution is contained in its confidence region. Thus the region corresponding to a 99% confidence level will be larger than that corresponding to a 95% confidence level: a larger confidence region is needed to increase the probability that the region contains the chance distribution. The confidence region method leads to a more subtle implementation of the Calibration norm: instead of calibrating degrees of belief to each dataset distribution (which is impossible when these marginal distributions conflict) one only needs to ensure that the belief function lies within the confidence region determined by each dataset (which is possible if one chooses a confidence level high enough to ensure that the regions do not conflict). Thus, instead of taking $\mathbb{E} = \{P : P(V_i) = freq_i(V_i)\}$, we take $\mathbb{E} = \{P : P(V_i) \in R_i(V_i)\}$. So in our example we have that $\mathbb{E} = \{P : P(X_1, X_2, X_3) \in R_1(X_1, X_2, X_3), P(X_2, X_3, X_4) \in R_2(X_2, X_3, X_4), P(X_1, X_2, X_4) \in R_3(X_1, X_2, X_4)\}$.

The confidence region approach also addresses the problem of small sample size. This is because the size of the confidence region depends on the number of individuals observed in the dataset and on the number of variables measured, as well as on the confidence level: fewer sampled individuals (or more measured variables) will lead to a wider confidence region and a less precise estimate of the data-generating distribution, ceteris paribus. Moreover, wider regions typically correspond to a more equivocal maximum entropy probability function selected by the Equivocation norm. By choosing a confidence level that is both high enough that the confidence regions can be taken to be plausible estimates of the marginal chance distributions and high enough that the confidence regions do not conflict, one simultaneously solves the problem of inconsistent marginal probabilities and specifies a probability function which is somewhat influenced by the dataset distributions, but not unduly so when the sample size is small.

Having considered an objective Bayesian response to the two challenges, let us consider the other two approaches that we have encountered so far—Pollock's approach and the similarity approach.

Pollock does not address the two challenges. This must be considered to be a further point against Pollock's approach.

We presented the similarity approach as a generic-probability approach. Although it is not subject to the problem of inconsistent marginals, the problem of small sample size still remains for the generic version of the similarity approach. We will now see that the similarity approach can be reconstructed as purely single-case, in order to circumvent the problem.

Suppose that raw data from different studies that investigate a certain disease is available. Certain variables measured by one of these studies may not be measured by another study. For example, one study on acute kidney disease may measure clinical variables (age, gender, co-morbidities etc.), another may measure pathology variables (creatine testing, proteinuria testing etc.) and a third may measure variables from imaging procedures. In addition, in each study, parts of the study results may be missing for some participants.

We can represent the situation as one of missing data. To avoid unnecessary complications, we focus on binary variables. We consider the set of all individuals who have participated in at least one study. For each of the participants the data will consist of an entry for all of the N variables that are at least measured in one study. The entries are either NM (not measured), if the person did not participate in a study where the variable has been measured or if the person did participate but no value has been recorded, 1 if the participant has the attribute expressed by the variable and 0 if the person did participate in the study but does not have the attribute expressed by the variable. Formally, the data consists of observations for n-individuals, $b_1, \ldots, b_n$. For each individual b_i, $i = 1, \ldots, n$, the data consists of a string of information $Dat_i = (x_{i,1} \ldots x_{i,N})$; where for all $j = 1 \ldots N$, $x_{i,j} \in \{0, 1, NM\}$ specifies the status of the individual b_i with respect to the attribute X_j. Suppose that for each individual b_i, $i = 1, \ldots, n$, measurements on A are available, i.e., $A(b_i) \in \{0, 1\}$.

Here is an example for 6 patients $b_1, \ldots, b_6$ and four variables in three datasets $V_1 = \{X_1, X_2, X_3\}, V_2 = \{X_2, X_3, X_4\}, V_3 = \{X_1, X_2, X_4\}$:

Patient	X_1	X_2	X_3	X_4
b_1	1	1	1	NM
b_2	0	1	0	NM
b_3	NM	1	1	0
b_4	NM	0	0	1
b_5	1	0	NM	1
b_6	0	0	NM	1

Gower's similarity measure for individuals, rather than for attributes, measures the number of shared attributes of b_k and b_l divided by the number of attributes where data for both b_k and b_l are available:

$$sim_{Gow}(b_k, b_l) = \frac{|\{j \in \{1, \ldots, N\} : X_j(b_k) = X_j(b_l)|}{|\{j \in \{1, \ldots, N\} : X_j(b_k) \neq NM \wedge X_j(b_l) \neq NM\}|} \quad (4.3)$$

The direct inference probability for Ac is a weighted average of the values of individuals in the database. The weighting is according to similarity.

Attribute similarity solution to the reference class problem on basis of raw data. The direct inference probability that c has attribute A is given by

$$P(Ac) = T \sum_{i=1}^{n} sim_{Gow}(b_i, c) A(b_i) \; , \quad (4.4)$$

where T is a normalising constant.

For instance, let Bc, Cc, Hc, Ec and

1. $Ab_1, \neg Eb_1, Cb_1$
2. $Ab_2, Bb_2, \neg Cb_2, Eb_2, Hb_2$
3. $\neg Ab_3, Bb_3, Cb_3$

then $Prob(Ac) = \frac{4}{7}(\frac{1}{2} \cdot 1 + \frac{3}{4} \cdot 1 + \frac{2}{2} \cdot 0) = \frac{5}{7}$.

This approach addresses both the above obstacles that face generic-probability approaches. In order to apply this method, one does not need a large sample of individuals who have a particular combination of attributes—we may assign a direct inference probability without considering frequencies in reference classes at all. Data in different datasets measuring different variables can be employed without worrying about the interaction of these variables; all the work is done by the similarity measure. Hence, no inconsistent marginals arise. Therefore, this single-case reinterpretation of the similarity approach apparently circumvents both the problem of small sample size and the problem of inconsistent marginal probabilities.

One might object to this apparent resolution as follows. Perhaps the single-case version of the similarity approach, although applicable, should not be applied when there is insufficient data, because, when there is insufficient data, it is not reasonable to infer anything about the individual. In response to this objection, note that the single-case version of the similarity approach is based on the assumption that the best way to determine the direct inference probability for a certain individual is to study individuals that have similar attributes. According to this approach, to which reference class these individuals belong is irrelevant; what is relevant is that they are similar to the individual in question. Therefore, data that is insufficient for estimating generic probabilities may yet be sufficient for direct inference. Indeed, single-case versions of the similarity approach have been successfully used in machine learning and medicine—see Jerez et al. (2010) and Davis et al. (2010).

4.6   A Mechanism-Based Approach

In this section, we identify two loci where evidence of mechanisms may be used to provide better solutions to the reference class problem. First, evidence of mechanisms may be used to enrich the problem formulation, to better capture the causal structure of the direct inference problem in question. Capturing more features of the problem promises to lead to more accurate direct inferences. Second, evidence of mechanisms may be used to provide information about reference class probabilities for which we have no statistical information available but which are relevant to the direct inference at hand. The method of *comparative process tracing* employs evidence of similarity of mechanisms to extrapolate frequencies from a reference class for which we have data to one that is relevant to the direct inference.

In this section, we explain the concept of mechanism and discuss these two situations in which evidence of mechanisms can help solve reference class conflicts.

4.6.1   Evidence of Mechanisms and Causal Structure

Illari and Williamson characterise mechanism in the following way: "A mechanism for a phenomenon consists of entities and activities organized in such a way that they are responsible for the phenomenon" (Illari and Williamson 2012, p. 120). Examples of mechanisms are the mechanism for drug metabolism in humans, the mechanism of natural selection and the mechanism of how supernovae arise. For instance, Russo and Williamson (2007, p. 162) describe the mechanism that leads from smoking to cancer by "The hair-like cilia in the lungs, which beat rhythmically to remove inhaled particles, are destroyed by smoke inhalation; thus the lung cannot cleanse itself effectively. Cancer-producing agents in cigarette smoke are therefore trapped in the mucus. Cancer then develops when these chemical agents alter the cells, in particular, cell division." According to Russo and Williamson, mechanisms play a crucial role in establishing causality. To establish that an event C causes an event E, normally two claims have to be established: that C and E are probabilistically dependent conditional on other causes of E, and that there exists a mechanism connecting C and E that can account for this correlation. One way of establishing that a certain mechanism connects C and E is to establish the crucial attributes of the mechanism, i.e., to establish that the crucial entities and activities are present and are organized in the right way.

How we can get evidence of a mechanism and what counts as high quality evidence of mechanism is an active area of research (for an overview see Clarke et al. 2014). Evidence of mechanisms can come from various sources, including laboratory experiments, literature reviews of basic science, expert testimony, confirmed theory or by analogy from, e.g., animal experiments (Clarke et al. 2014).

In what is to follow, instead of speaking of a single mechanism, we are going to speak of the *mechanistic structure* that gives rise to an attribute measured in a reference class. The mechanistic structure consists of all the mechanisms that explain the attribute, either by inducing the attribute or by inhibiting it or by moderating its value (i.e., by changing its value or limiting change that would otherwise occur).

Evidence of mechanisms can assist direct inference by providing information about causal structure. As we have seen, in direct inference we seek to ascertain the probability that a particular individual instantiates a particular attribute. Statistical information is normally available, which takes the form of the generic probability of the attribute of interest conditional on other attributes which define a reference class. It is usually the case that there is also information to hand about the mechanisms that give rise to the attribute of interest and that connect this attribute to those that define the reference class. That this information about mechanisms is often qualitative rather than quantitative does not, of course, imply that it should be ignored for the purposes of direct inference. Evidence of mechanisms can be taken into account by helping to ascertain the causal structure that connects the attribute of interest to those attributes that characterise the reference classes for which we have statistics.

To the extent that this causal and statistical information underdetermines the required direct inference probability, one can apply objective Bayesian methods to select a direct inference probability that satisfies the constraints imposed by the available evidence but which is otherwise equivocal.

Constraints on the objective Bayesian probability function arising from the available statistical information can be dealt with by the approach introduced in Sect. 4.2. But now there are also causal constraints, which arise from evidence of mechanisms in combination with available statistical information. Williamson (2005a, §5.8) describes how causal constraints can be taken into account by the objective Bayesian approach. Briefly, objective Bayesianism adopts the principle that, when one learns of new variables which are not causes of the old variables, probabilities over the old variable set should not change. This principle allows one to translate information about causal relationships into constraints that equate probabilities. One can then identify the probability function $P^\dagger$ with maximum entropy, from all those probability functions which satisfy the constraints arising from statistical information together with the equality constraints arising from causal information. As described in Sect. 4.2, $P^\dagger$ is used for direct inference.

The mechanistic approach involves enriching the problem formulation by taking causal structure and extra causes and effects into account. The question arises, then, as to whether there are methods for mitigating this extra complexity. Maximising entropy is a computationally demanding optimisation problem, and practical methods are required to ensure that the optimisation can be carried out. Bayesian networks are often used to reduce the complexity of probabilistic inference, and they can also be applied here, as we shall now explain.

The probability function $P^\dagger$ advocated by objective Bayesianism can be represented by a Bayesian network—this is called an *objective Bayesian net* (Williamson 2005b; Landes and Williamson 2016). A Bayesian net consists of a directed acyclic graph whose nodes are variables together with the probability distribution of each variable conditional on its parents in the graph (Pearl 1988). The *Markov condition*, which says that each variable is probabilistically independent of its non-descendants in the graph conditional on its parents, then enables the Bayesian net to completely determine a joint probability distribution over all the variables that occur in the net. It turns out that, in virtue of its appeal to maximum entropy methods, the objective Bayesian probability function $P^\dagger$ typically satisfies many probabilistic independence relationships and these relationships can be identifiable in advance, to build an objective Bayesian net representation of $P^\dagger$ (Williamson 2005a, §§5.6–5.8). The advantages of the Bayesian net representation are that (i) it is a more economical way to specify a probability function than simply specifying the probability of each combination of values of the variables under consideration, and (ii) a whole host of efficient inference algorithms have been developed to calculate conditional probabilities—such as are required for direct inference—from the network.

Note that in general, the arrows in the graph of a Bayesian net are merely a formal device to represent certain probabilistic independencies—they would not normally be interpretable as representing causal relationships or other sorts of relationship. However, it turns out that in many situations where causal information is available, the objective Bayesian net is interpretable as a *causal* Bayesian net, i.e., the arrows in the directed acyclic graph represent causal relationships amongst the variables. This is so, for example, when the evidence consists of the causal structure and constraints on the probability distribution of each variable conditional on its parents; then the maximum entropy probability function $P^\dagger$ that is advocated by objective Bayesianism is determined by a causal net involving that causal structure—i.e., the Markov condition provably holds (Williamson 2005a, Theorem 5.8). So, in many situations the objective Bayesian approach can be thought of as a causal network approach.

In sum, the mechanistic approach to the reference class problem motivates two developments to the objective Bayesian approach. First, causal constraints need to be taken into account, in addition to probabilistic constraints. Second, Bayesian networks can be used to make direct inference more computationally tractable.

Let us consider an example. Bernd is a white man from the US who has never been to hospital. Bernd smokes and is highly physically active. We are interested in whether Bernd will get a stroke (St). Bernd belongs to two reference classes for which we have data: he smokes (S), and he is physically active (A). On the one hand, smoking increases the risk of getting a stroke by a half, $P^*(St|S) = 1.5 \cdot P^*(St|\neg S)$ (Shinton and Beevers 1989). On the other hand, a high degree of physical activity decreases risk of stroke compared to a low degree of physical activity, $P^*(St|A) = 0.79 \cdot P^*(St|\neg A)$ (Mozaffarian et al. 2015). Belonging to the two reference classes yields opposite conclusions as to whether Bernd will get a stroke. What is the direct inference probability that Bernd will get a stroke? Mechanistic evidence is useful to constrain the direct inference probability and for an improved understanding of the causal structure of the problem. An improved understanding of the causal structure can yield tighter constraints on the direct inference probabilities, and can thus lead to direct inference probabilities that are better calibrated to the true chances.

The mechanism-based approach proceeds as follows. As a first step, we use available information about the mechanisms for stroke that involve smoking and physical activity and use this information to construct a causal net which schematically represents the causal relationships connecting these variables. In examples such as this, there is usually plenty of information about mechanisms available in the literature. Indeed, there are many mechanisms connecting smoking to stroke. Smoking increases the risk of building blood clots and blood clots increase the risk of stroke. Smoking increases the risk of high blood cholesterol (C) and high cholesterol increases the risk of some kinds of stroke. Similarly, there are many inhibiting mechanisms connecting physical activity to stroke. Physical activity reduces obesity and obesity is a major risk factor for stroke. Physical activity prevents hypertension, high cholesterol and the development of blood clots. Hypertension increases the risk of stroke.

The aim of the present example is to illustrate the mechanism-based approach, rather than to provide a detailed analysis of all mechanisms relating stroke, physical activity, and smoking. We therefore simplify the example by taking the mechanism involving high cholesterol to be the only mechanism that is influenced by both smoking and physical activity. Clearly, the example can be further enriched to take into account other evidence of mechanisms, and to lead to further improvements in direct inference.

Employing mechanistic evidence translates into the causal structure depicted in Fig. 4.1. Both smoking and physical activity influence cholesterol levels. Therefore, we draw an arrow from S to C and from A to C in the causal graph. Since high cholesterol causes stroke, we draw an arrow from C to St. Smoking and physical activity influence stroke via at least two different non-overlapping mechanisms. Therefore, we draw arrows from both S and A to St.

As a second step, we need to ascertain any other available probabilities that are relevant to the variables that occur in the causal graph. These probabilities can often

Fig. 4.1 Causal graph for the stroke example

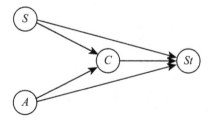

also be found in the literature. In order to determine $P^*(St|S \wedge A)$, it would suffice to specify, for every variable appearing in the graph, the probability distribution of that variable conditional on its parents in the graph. This would constitute a causal Bayesian net, which would fully determine the joint chance distribution over the variables that occur in the graph.

However, it will typically be the case that not all of these probabilities are available in the literature. Instead, the conditional probability of a variable given some of its parents might be available. In this case, we treat these "incomplete distributions" as constraints and carry out maximum entropy direct inference, as described above.

In the literature, relative risks are often reported. For instance, $P^*(St|S) = 1.5 \cdot P^*(St|\neg S)$ (Shinton and Beevers 1989). In this case, we obtain the absolute risk of stroke given smoking $P^*(St|S)$, if we know the base rate or prevalence of stroke and hypertension, $P^*(St)$ and $P^*(S)$:

$$P^*(St) = P^*(St|S)P^*(S) + P^*(St|\neg S)(1 - P^*(S))$$
$$= 1.5 \cdot P^*(St|\neg S)P^*(S) + P^*(St|\neg S)(1 - P^*(S))$$

From the literature, we obtain the prevalence or base rate of stroke $P^*(St) = 0.027$, of smoking $P^*(S) = 0.168$, of high cholesterol $P^*(C) = 0.131$ and of physical activity $P^*(A) = 0.695$ (Mozaffarian et al. 2015). Hence, $P^*(St|S) = 0.032$ and $P^*(St|A) = 0.025$.

Estimates of the risk of stroke given high cholesterol compared to the case of no high cholesterol are more controversial:

> The role of blood cholesterol in stroke prevention is unclear. Most prospective studies have failed to find a relation between total cholesterol and risk of total stroke. It has been proposed that this may be due to the differing association with subtypes of stroke. An inverse association has been observed with hemorrhagic strokes and a positive association with ischemic stroke. (Wannamethee et al. 2000, p. 1887)

We differentiate between ischemic stroke ($St = 1$) and haemorrhagic stroke ($St = 2$). We abbreviate $St = 1 \vee St = 2$ by St. The following estimates can be found in the literature: $P^*(St = 1|C) = 1.4 \cdot P^*(St = 1|\neg C)$ (Benfante et al. 1994) and $P^*(St = 2|C) = 0.69 \cdot P^*(St = 2|\neg C)$ (Wang et al. 2013). The direct inference probability can be further constrained by, for instance, ascertaining $P^*(C|S)$, $P^*(C|A)$ and the high cholesterol-smoking interaction with respect to the development of stroke $P^*(St|C \wedge S)$.

As a third step, we can apply objective Bayesian methods to infer a direct inference probability. This probability is determined by the probability function with maximum entropy, from all those that satisfy the constraints imposed by the causal and statistical information, as explained above. This way, we obtain the direct inference probability that is compatible with the available evidence, but otherwise equivocal.

4.6.2   Evidence of Mechanisms and Extrapolation

Evidence of mechanisms can also be used to help extrapolate available reference class probabilities to classes for which we have no statistical information available but which are relevant to the direct inference at hand. Thus, a direct inference probability can be influenced by frequencies obtained from reference classes to which the individual does not belong.

Broadly speaking, in order to extrapolate a probabilistic claim from one reference class to another, one needs to show that the determinants of the probability in the target class are sufficiently similar to those in the source class. One can do this by showing that the underlying mechanisms, contextual factors and background conditions are sufficiently similar in the two classes. Clearly, evidence of mechanisms is crucial to this mode of inference.

There are various strategies of employing evidence of mechanisms to assist extrapolation (Parkkinen and Williamson 2017). *Comparative process tracing* is one such strategy (Steel 2008). To determine how likely an extrapolation from a model organism to a target organism is to succeed, Steel proposes that one has to learn the mechanism in the model organism and that one has to compare stages of the mechanism in which the mechanism in the model and the target are most likely to differ significantly. He then concludes that "in general, the greater the similarity of configuration and behavior of entities involved in the mechanism at these key stages, the stronger the basis for the extrapolation" (Steel 2008, p. 89). Steel illustrates his method by means of an example. According to Steel, rats are better models than mice for determining whether Aflatoxin B1 causes liver cancer in humans. He argues that (i) phase 1 and phase 2 metabolism are the crucial parts in most carcinogenic mechanisms among mammals and that (ii) the phase 1 metabolism is similar in all three species and that (iii) there are important similarities between rats and humans in phase 2 metabolism but dissimilarities between humans and mice. Therefore, according to comparative process tracing, the extrapolation from rats to humans is more likely to succeed.

Consider now the prevalence of smoking in a country, state or city. Can we extrapolate the chance to another country, state or city? This depends on how similar the crucial attributes or determinants of the smoking behaviour are in the target and the study population. For instance, demographic or socioeconomic factors are important determinants of smoking behaviour: e.g., education, income level of the country, age, and gender (Hosseinpoor et al. 2011). It is legitimate to extrapolate the prevalence of smoking from one state in a high-income country to another state, provided that there is little difference in socioeconomic or demographic factors. For instance, in Austria, it is reasonable to extrapolate the chance of smoking from the State of Tirol to the State of Lower Austria. Indeed, the smoking rates in Austria in 7 out of 9 states differ by less than 2% (20.9%–22.7%) (Statistik Austria 2016). It is, however, not reasonable to extrapolate from the State of Tirol to the State of Vienna. While the State of Vienna consists roughly of the large city, the State of Tirol consists mainly of rural areas and smaller cities.

Often it is impossible to avoid extrapolation in direct inference, because statistics in the relevant reference classes are not available. Thus in the stroke example, in order to specify probabilities relevant to the causal net, we extrapolated estimates for the probabilities in the study population to the population to which Bernd belongs. For instance, since Bernd has never been hospitalised, Bernd does not belong to any of the hypercholesterol populations from which the samples were drawn—these studies were conducted on hospital patients. Less straightforward extrapolation has been carried out to obtain the risk of ischemic stroke given high blood cholesterol. The estimate $P^*(St = 1|C) = 1.4 \cdot P^*(St = 1|\neg C)$ from Benfante et al. (1994) is obtained from studies conducted on Hawaiian Japanese men. There are surely important differences between Hawaiian Japanese men and US-born citizens. However, "the associations of major risk factors with CHD and stroke were very similar to those found for US white men" (Benfante et al. 1994, p. 818). This provides grounds that an extrapolation from the risk of ischemic stroke to Bernd's population (US white men) will be successful, i.e., that $P^*(St = 1|C) \approx 1.4 \cdot P^*(St = 1|\neg C)$ in US white men.

4.7   Conclusions

We have presented four approaches to the reference class problem. The objective Bayesian approach advocates degrees of belief that are compatible with the evidence and otherwise equivocal. Pollock's approach is based on combinatorial probabilities. The similarity approach is based on similarity of attributes or of individuals. Finally, the mechanistic approach allows one to enrich and refine the problem formulation, in order to achieve more credible direct inferences. Which of the approaches should we prefer?

To give an answer, at least three further questions have to be considered. First, how is the aggregation of reference classes done? The similarity approach identifies the direct inference probability with a weighted average of reference class frequencies. This weighted average seems to be somewhat arbitrary: why should we combine different evidence exactly in this way? The objective Bayesian approach and Pollock's approach relate the direct inference to less ad hoc quantities: respectively, the maximum entropy probability for the narrowest reference class to which the individual is known to belong, and the expectable value of the narrowest reference class. We argued above, however, that Pollock's justification of the claim that the expectable value is often close to the true chance is unconvincing. Advocating equivocal degrees of belief on the basis of good but incomplete evidence is preferable to advocating degrees of belief on the basis of complete but poor evidence. Hence, aggregation according to the objective Bayesian approach might be considered preferable to aggregation via the other two approaches.

Second, what kind of evidence can be used? The attribute-based similarity measure does not account for causal similarity. Generally, measures based on the number of shared attributes can only be seen as a first approximation to a similarity measure that will be useful for direct inference. The objective Bayesian account can account for causal information by exploiting evidence of mechanisms. This leads to tighter empirical constraints on the direct inference probability. Of course, that it is able to exploit evidence of mechanisms is only an advantage if there is evidence of mechanisms available that is of sufficiently high quality.

Third, the similarity approach can be reconstructed as a single-case-probability approach and may be used even when the sample size is too small to reliably estimate the data-generating chances. The strategy of taking into account similarity of individuals might thus compensate for low sample size.

Our suggestion is to prefer the similarity approach where sample sizes are too small to sensibly apply generic-probability approaches. In the other cases, the objective Bayesian approach is to be preferred. This is especially true if evidence of mechanisms is available that constrains the direct inference probability to lie in a tight interval. In such a case, there is less of a role for the Equivocation norm and most of the work is being done by the Calibration norm. This is likely to lead to accurate, rather than cautious, direct inference probabilities.

Acknowledgements The research for this paper has been funded by the Arts and Humanities Research Council via grant AH/M005917/1.

References

Benfante, R., K. Yano, L.-J. Hwang, J.D. Curb, A. Kagan, and W. Ross. 1994. Elevated serum cholesterol is a risk factor for both coronary heart disease and thromboembolic stroke in hawaiian Japanese men. Implications of shared risk. *Stroke* 25(4): 814–820.

Clarke, B., D. Gillies, P. Illari, F. Russo, and J. Williamson. 2014. Mechanisms and the evidence hierarchy. *Topoi* 33(2): 339–360.

Davis, D.A., N.V. Chawla, N.A. Christakis, and A.-L. Barabási. 2010. Time to care: A collaborative engine for practical disease prediction. *Data Mining and Knowledge Discovery* 20(3): 388–415.

Gower, J.C. 1971. A general coefficient of similarity and some of its properties. *Biometrics* 27(4): 857–871.

Haybittle, J., R. Blamey, C. Elston, J. Johnson, P. Doyle, F. Campbell, R. Nicholson, and K. Griffiths. 1982. A prognostic index in primary breast cancer. *British Journal of Cancer* 45(3): 361.

Hosseinpoor, A.R., L.A. Parker, E.T. d'Espaignet, and S. Chatterji. 2011. Social determinants of smoking in low-and middle-income countries: Results from the world health survey. *PLoS One* 6(5): e20331.

Illari, P.M., and J. Williamson. 2012. What is a mechanism? Thinking about mechanisms *across* the sciences. *European Journal for Philosophy of Science* 2: 119–135.

Jaynes, E.T. 1957. Information theory and statistical mechanics. *The Physical Review* 106(4): 620–630.

Jerez, J.M., I. Molina, P.J. García-Laencina, E. Alba, N. Ribelles, M. Martín, and L. Franco. 2010. Missing data imputation using statistical and machine learning methods in a real breast cancer problem. *Artificial intelligence in medicine* 50(2): 105–115.

Landes, J., and J. Williamson. (2016). Objective Bayesian nets from consistent datasets. In *Proceedings of the 35th International Workshop on Bayesian Inference and Maximum Entropy Methods in Science and Engineering*, ed. A. Giffin, and K.H. Knuth, vol. 1757. American Institute of Physics Conference Proceedings, Potsdam

Mozaffarian, D., E.J. Benjamin, A.S. Go, D.K. Arnett, M.J. Blaha, M. Cushman, S. de Ferranti, J.-P. Despres, H.J. Fullerton, V.J. Howard, et al. 2015. Heart disease and stroke statistics-2015 update. A report from the American Heart Association. *Circulation* 131(4): E29–E322.

Parkkinen, V.-P., and J. Williamson. 2017. Extrapolating from model organisms in pharmacology. In *Uncertainty in pharmacology: Epistemology, methods, and decisions*, ed. B. Osimani. Dordrecht: Springer.

Pearl, J. 1988. *Probabilistic reasoning in intelligent systems: Networks of plausible inference*. San Mateo: Morgan Kaufmann.

Pollock, J.L. 2011. Reasoning defeasibly about probabilities. *Synthese* 181(2): 317–352.

Pollock, J.L. 2016. LISP code for OSCAR. http://johnpollock.us/ftp/OSCAR-web-page/oscar. html. Accessed 19 Sept 2016.

Russo, F., and J. Williamson 2007. Interpreting causality in the health sciences. *International Studies in the Philosophy of Science* 21(2): 157–170.

Shinton, R., and G. Beevers. 1989. Meta-analysis of relation between cigarette smoking and stroke. *BMJ* 298(6676): 789–794.

Statistik, Austria. 2016. Smoking statistics in Austria 2014. http://www.statistik.at/web_de/ statistiken/menschen_und_gesellschaft/gesundheit/gesundheitsdeterminanten/rauchen/index. html. Accessed 28 Sept 2016.

Steel, D. 2008. *Across the boundaries: Extrapolation in biology and social science*. Oxford/New York: Oxford University Press.

Venn, J. 1888. *The logic of chance*, 3rd ed. London: Macmillan.

Wallmann, C. 2017. A Bayesian solution to the conflict of narrowness and precision in direct inference. *Journal for General Philosophy of Science*.

Wallmann, C., and G.D. Kleiter 2014a. Degradation in probability logic: When more information leads to less precise conclusions. *Kybernetika* 50(2): 268–283.

Wallmann, C., and G.D. Kleiter 2014b. Probability propagation in generalized inference forms. *Studia Logica* 102(4): 913–929.

Wang, X., Y. Dong, X. Qi, C. Huang, and L. Hou. 2013. Cholesterol levels and risk of hemorrhagic stroke. A systematic review and meta-analysis. *Stroke* 44(7): 1833–1839.

Wannamethee, S.G., A.G. Shaper, and S. Ebrahim. 2000. HDL-cholesterol, total cholesterol, and the risk of stroke in middle-aged British men. *Stroke* 31(8): 1882–1888.

Williamson, J. 2005a. *Bayesian nets and causality: Philosophical and computational foundations*. Oxford: Oxford University Press.

Williamson, J. 2005b. Objective Bayesian nets. In *We will show them! essays in Honour of Dov Gabbay*, ed. S. Artemov, H. Barringer, A.S. d'Avila Garcez, L.C. Lamb, and J. Woods, vol. 2, 713–730. London: College Publications.

Williamson, J. 2010. *In defence of objective Bayesianism*. Oxford: Oxford University Press.

Williamson, J. 2013. Why frequentists and Bayesians need each other. *Erkenntnis* 78(2): 293–318.

Williamson, J. 2017a. *Lectures on inductive logic*. Oxford: Oxford University Press.

Williamson, J. 2017b. Models in systems medicine. *Disputatio*. In press.

Part II
Structures for Quantum Experiments

Chapter 5
Separate Common Causes and EPR Correlations: An "Almost No-Go" Result

Leszek Wroński, Tomasz Placek, and Michał Tomasz Godziszewski

Abstract One diagnosis of Bell's theorem is that its premise of Outcome Independence is unreasonably strong, as it postulates one common screener system that purports to explain all the correlations involved. This poses a challenge of constructing a model for quantum correlations that is local, non-conspiratorial, and has many separate screener systems rather than one common screener system. In particular, the assumptions of such models should not entail Bell's inequalities. Although we stop short of proving that such models exist (or do not exist), we obtain a few results relating them to models with a common screener system. We prove that if such a model exists, then there exists a local common screener system model for quantum correlations breaking Bell's inequalities; that model, however, can be conspirational. We also show a way of transforming a model with separate screener systems for the Bell-Aspect correlations that satisfies strong Parameter Independence (PI) and No-Conspiracy (NOCONS) into a somewhat different model for the same correlations in which strong PI is somewhat compromised, but NOCONS and "regular" PI hold, and the elements of the constructed partition act deterministically with respect to measurement results. This means that such models allow the derivation of the Clauser-Horne-Shimony-Holt inequalities.

5.1 Introduction

Bell's (1964) theorem derives a testable probabilistic inequality from the assumption that quantum mechanics can be completed by states more informative than quantum states and that these "hidden" states satisfy some intuitive assumptions. Since the inequality is violated by quantum mechanical predictions, and over the years a consensus has grown that it is also empirically violated, at least one of the premises of the derivation must be false. Accordingly, Bell's paper poses a challenge of

L. Wroński (✉) • T. Placek
Jagiellonian University, Kraków, Poland
e-mail: leszek.wronski@uj.edu.pl

M.T. Godziszewski
Logic Department, Institute of Philosophy, University of Warsaw, Warsaw, Poland

© Springer International Publishing AG 2017
G. Hofer-Szabó, L. Wroński (eds.), *Making it Formally Explicit*, European Studies in Philosophy of Science 6, DOI 10.1007/978-3-319-55486-0_5

explaining the theorem, that is, arguing which of its premises is false and why. It is this project that we want to contribute to in this paper.

In this paper we will focus on a later and more advanced version of Bell's theorem that assumes a probabilistic working of the hidden states in bringing about the outcomes of measurements. Two premises of the derivation draw on the idea of locality, which says that an event can be influenced by a remote event only by a mediation of neighboring (local) events. Since in the setup relevant to Bell's theorems, outcomes registered in one wing of the experiment are space-like separated from the selection of settings ("parameters") made in the other wing of the experiment, this idea leads to two independence conditions, typically called "Parameter (or Context) Independence" (PI) and "Outcome Independence" (OI). PI says that, given the hidden state, outcomes registered at a nearby measurement apparatus are independent from settings selected at a remote measurement apparatus. OI requires that, given the hidden state, outcomes registered at a nearby apparatus are independent from outcomes registered at the remote apparatus. The remaining premise of the derivation, called "No Conspiracy" (NOCONS), postulates that selections of measurement settings are free, that is, independent from the hidden states.

Following a large part of literature, by "independent" we will mean here "probabilistically independent". Accordingly, the three conditions of PI, OI, and NOCONS are given here this reading: (PI) Given each hidden state, outcomes registered at a nearby measurement apparatus are probabilistically independent from settings selected at a remote measurement apparatus; (OI) Given each hidden state, outcomes registered at a nearby apparatus are probabilistically independent from outcomes registered at a remote apparatus; (NOCONS) selections of measurement settings are probabilistically independent from the hidden states.

A diagnosis of Bell's theorem that we analyse here (and argue against) stems from observing a certain subtlety in OI (the observation was first made by Belnap and Szabó (1996)). OI bears an affinity to what is known as Reichenbach's (1956) screening-off condition, which concerns a pair of correlated events and a third event "screening off" one of the events belonging to the pair from the other (for a rigorous formulation see below). The difference that the above researchers noticed is that in the context of Bell's theorem, OI posits a single screening factor for *many* pairs of correlated results, and hence seems to be unreasonably strong.

This "single vs. many" dialectics motivates a project of deriving Bell's inequalities from weaker premises, with Outcome Independence being replaced by its more modest relative. A hope was that with the new premise, Bell's inequalities could not be derived, which would put the blame for the derivation on the "old" Outcome Independence.

This poses a challenge of constructing a model for quantum correlations that is local, non-conspiratorial, and has many separate screener systems rather than one common screener system. In particular, the assumptions of such models should not entail the Bell inequalities. We prove that if a model described exists, then there exists a local common screener system model (possibly conspiratorial) for quantum correlations breaking Bell's inequalities. Additionally we present the state of research concerning the project of proving the non-existence of local, non-conspiratorial separate-ss models; our theorem uses the notion of "strong Parameter Independence" introduced in Sect. 5.3.1.

Our paper is organized as follows: The next section sketches the background of the project we criticize. Section 5.3 gives formal definitions pertaining to the distinction we alluded to. With these definitions in hand, in Sect. 5.4 we offer a survey of earlier results. The main Sects. 5.5 and 5.6 contain our results (including detailed proofs); the paper ends with Conclusions.

5.2 Background: From Bell's Local Causality to Separate Systems of Screeners

Bell proved a stochastic version of his theorem from a premise he called "local causality".[1] As he explained (1975), the underlying idea is that if 1 and 2 are space-like separated regions, then events occurring in 1 should not be causes of events occurring in region 2. Such events could be correlated, he acknowledged, as they might have a common cause. Moreover, they may remain to be correlated, even if the probability is conditioned on a specification Λ of the state of the events' common past, i.e., $P(A|\Lambda B) \neq P(A|\Lambda)$, where A and B stand for events occurring in regions 1 and 2, respectively. However, he claims that "in the particular case that Λ contains already a complete specification of beables in the overlap of the two light cones, supplementary information from region 2 could reasonably be expected to be redundant [for probabilities of events in region 1]", which he takes for justification of this screening-off formula[2]:

$$P(A|\Lambda B) = P(A|\Lambda).^3$$

Accordingly, Λ represents here a full specification of the state in the common past. The formula is tacitly universally quantified, that is, it should read "for every possible state in region ..., if Λ is its full specification, then ...".

Almost two decades earlier, (Reichenbach 1956, p. 159) hit upon a similar idea, while attempting to analyze the arrow of time in terms of causal forks:

> In order to explain the coincidence of A and B, which has a probability exceeding that of chance coincidence, we assume that there exists a common cause C. [...] We will now introduce the assumption that the fork ACB satisfies the following relations:
>
> $$P(AB|C) = P(A|C)P(B|C) \quad P(AB|\neg C) = P(A|\neg C)P(B|\neg C)$$
> $$P(A|C) > P(A|\neg C) \qquad P(B|C) > P(B|\neg C).$$

---

[1] The stochastic version assumed probabilistic working of hidden states; there were other premises of the proof.

[2] At this point in his paper Bell changes notation, introducing variables for states in each event's backward light-cone. For a recent analysis of Bell's local causality, cf. Norsen (2006) and Seevinck and Uffink (2011).

[3] We frequently omit the "∩" sign between names of events.

The two formulas on the top are called the (positive and negative) screening-off conditions, whereas the two at the bottom are known as the conditions of positive statistical relevance. It is easy to note the same motivation behind both Bell's causal locality and Reichenbach's screening-off condition. Since $P(AB|C) = P(A|C)P(B|C)$ is equivalent to $P(A|BC) = P(A|C)$ if $P(C) \neq 0$, the two concepts are formally similar as well, though the former allows for any number of factors ("screeners") to be conditioned upon, whereas the latter is dichotomous, since it admits as screeners an event and its negation only.

In the 1970s and 1980s Reichenbach's project was continued by W. Salmon. It was most likely van Fraassen (1982) who first saw the connection between Bell's local causality and the screening-off condition, which, generalized to any number of screeners and applied to pairs of outcomes, with screeners identified with hidden states, is just Outcome Independence. But this condition—taken together with two more premises—entails Bell's inequalities, which are both violated by quantum mechanics and most likely experimentally falsified. Since the two other premises look intuitive, a popular diagnosis was to reject OI. This means, however, to reject the screening-off condition, generalized from dichotomous screeners to any number of screeners. But since the screening-off condition (so generalized) may appear to be a mathematical tautology,[4] how could it be empirically falsified?

A diagnosis that seems to resolve the conflict came from Belnap and Szabó's (1996) distinction between common causes and *common* common causes. Observe that in Reichenbach's approach one posits a system of screeners, i.e., C and $\neg C$, for a *single* pair of events. In contrast, in the context of Bell's theorem, one envisages a large number of correlated pairs of results, produced in mutually exclusive measurements (i.e., represented by non-commuting observables). Bell's local causality postulates a single set of screeners (full specifications of states in a relevant region) for all these correlations. The set might be arbitrarily large, but, importantly, each element of it is supposed to apply to *all* correlated pairs, making the events independent, conditional on each screener. In the recent terminology of Hofer-Szabó (2008), standard Bell-type theorems assume a *common* screener system, i.e, every element of this system pertains to all correlations under consideration—in contrast with *separate* screener systems.

With this distinction at hand, it is tempting to believe that, while the screening-off condition (as applicable to a single correlation) is correct, what lands us in trouble in the context of Bell's theorem is its extension which requires a common system of screeners for all the correlations. To put it differently, moving from a common screener system to separate screener systems relaxes Outcome Independence and Parameter Independence. The usual OI postulates that every correlation considered is screened off by every factor from the common screener system. The modified

---

[4]Especially since the following simple fact holds: let $\langle \Omega, \mathcal{F}, P \rangle$ be a probability space. Let $\mathcal{G} = \{\{A_i, B_i\}\}_{i \in I} \subseteq \mathcal{F}^2$ be a finite family of pairs of correlated events in $\mathcal{F}$. Then there exists a partition $\mathcal{C}$ of Ω such that for any $C \in \mathcal{C}$ and for any $i \in I$, $P(A_i B_i|C) = P(A_i|C)P(B_i|C)$. To construct this partition simply take all Boolean combinations of all correlated events, throwing out the empty ones should they arise.

condition, call it OI', requires that every correlation is screened off by every factor from the screener system *for this correlation* only. A similar change affects Parameter Independence.

To justify or reject this belief, starting from work reported in Szabó (2000), researchers have attempted to construct models of Bell-type correlations that would assume the existence of separate screener systems, satisfy the weakened premises of Parameter Independence (PI') and Outcome Independence (OI'), meet the No Conspiracy requirement, and would *not* be committed to Bell-type inequalities.[5,6] We prove that the if a model described exists, then there exists a local common screener system model for quantum correlations breaking Bell's inequalities. Such a model, however, is necessarily conspirational.

5.3 Screener Systems: Definitions

In the late 1990s the "Budapest school" of M. Rédei, L. Szabó, G. Hofer-Szabó, B. Gyenis and others, building upon the "*common* common causes vs. common causes" distinction, launched two projects[7]—to be briefly stated as below:

1. Is Reichenbach's common cause principle—or its generalization in form of some common common cause principle—tenable?
2. Are there models for Bell's correlations which are local, non-conspiratorial, have separate screener systems for each correlation, and are not committed to Bell-type inequalities?

The projects are different: the first must pertain to (generalizations of) the positive statistical relevance conditions, which are not required by the models of the Bell-Aspect correlations.[8] In turn, Bell's theorem brings in the issues of locality and no conspiracy which are not present in the discussions of common causes. Despite these differences, the method of handling these questions is the same in the Budapest school, and it boils down to asking if probability spaces respecting certain probabilistic constraints exist.

We are here concerned with the second question and thus the models we discuss are probability spaces, constrained by some conditions which are supposed to capture the spatiotemporal aspect inherent in locality as well as the modal aspects inherent in the conditions of no conspiracy and non-commutativity of quantum

[5]In some papers following Szabó's work these conditions are called Locality, No Conspiracy, and Screening-off, resp.

[6] Szabó paper reports on his computer simulations aimed to construct such models.

[7]Cf. Hofer-Szabó et al. (1999) and Szabó (2000).

[8]By "Bell-Aspect experiment" we mean the experiment envisioned by Bell (1964) and famously carried out by Aspect et al. (1982), the essence of which we recall below. The correlations investigated by Aspect et al. we also call Bell-Aspect correlations. These correlations violate the so-called Clauser-Horne-Shimony-Holt inequalities.

observables. We will introduce these constraints in turn. First let us recall the definition of a (classical) probability space.

Definition 1 (probability space) A *probability space* is a triple $\langle \Omega, \mathcal{F}, P \rangle$ such that:

- Ω is a non-empty set (sometimes called 'sample space');
- $\mathcal{F}$ is a σ-algebra of subsets of Ω (sometimes called 'algebra of events');
- P is a function from $\mathcal{F}$ to $[0, 1] \subseteq \mathbb{R}$ such that

  - $P(\Omega) = 1$;
  - P is countably additive: for a countable family $\mathcal{G}$ of pairwise disjoint elements of $\mathcal{F}$, $P(\cup \mathcal{G}) = \sum_{A \in \mathcal{G}} P(A)$.

P is called the *probability function* (or *measure*).

During our argument we begin with a probability space modelling the Bell-Aspect experiment. Then we construct a chain of transformations of the space, such that the probabilities of the events representing the measurement results and detector settings are preserved under them and the "fine-grained" space we end up with has interesting properties regarding screening-off (see Lemma 1, p. 98).

Now let us introduce the concept of a screener system.

Definition 2 (screener system) Let $\langle \Omega, \mathcal{F}, P \rangle$ be a probability space and $A, B \in \mathcal{F}$. A partition $\{C_i\}_{i \in I}$ of Ω satisfying for any $i \in I$

$$P(AB|C_i) = P(A|C_i)P(B|C_i) \tag{5.1}$$

is called a screener system for $\langle A, B \rangle$.[9]

Since the sum of the probabilities of the elements of a screener system equals 1, only countably many of them can be positive (see e.g. Theorem 10.2 in Billingsley (1995)). And so, while screener systems may be infinite, they are at most countably infinite, otherwise some conditional probabilities would not be defined.

A straightforward calculation proves the following fact:

Fact 3 *If* $\{C_i\}_{i \in I}$ *is a screener system for* $\langle A, B \rangle$, *then it is also a screener system for each of the following pairs:* $\langle A, B^{\perp} \rangle$, $\langle A^{\perp}, B \rangle$, *and* $\langle A^{\perp}, B^{\perp} \rangle$, *where* $X^{\perp} = \Omega \setminus X$.

This fact notwithstanding, the above definition leaves open how it should be applied to many pairs of events, if these pairs are not algebraic combinations of one another, as displayed above. As an example, consider two pairs, $\langle A, B \rangle$ and $\langle D, E \rangle$ such that $\langle D, E \rangle$ is identical to none of the following pairs: $\langle A, B^{\perp} \rangle$, $\langle A^{\perp}, B \rangle$, $\langle A^{\perp}, B^{\perp} \rangle$ $(A, B, D, E \in \mathcal{F})$. We may then postulate two separate systems of

---

[9]This is in essence definition 5 of a screener-off system of Hofer-Szabó (2008); in contrast to his definition, the screener system is defined here for a pair of events, regardless of whether or not they are correlated. In the sequel we frequently omit the brackets when speaking about correlated pairs of events.

screeners, $\{C_{AB}^k\}_{k<K(AB)}$[10] and $\{C_{DE}^{k'}\}_{k'<K(DE)}$, one for $\langle A, B\rangle$ and the other for $\langle D, E\rangle$, that is: for every $k < K(AB)$ and $k' < K(DE)$,

$$P(AB|C_{AB}^k) = P(A|C_{AB}^k)P(B|C_{AB}^k) \text{ and } P(DE|C_{DE}^{k'}) = P(D|C_{DE}^{k'})P(E|C_{DE}^{k'}).$$

Alternatively we may postulate a single common screener system $\{C_i\}_{i \in I}$ for the two pairs satisfying for every $i \in I$:

$$P(AB|C_i) = P(A|C_i)P(B|C_i) \text{ and } P(DE|C_i) = P(D|C_i)P(E|C_i).$$

To rigorously introduce the weakened versions of Outcome Independence (OI'), Parameter Independence (PI'), and No Conspiracy (NOCONS), let us recall the setup of Bell's theorem. A source emits pairs of objects, and the members of each pair travel in separate "wings" of the experiment towards remote detectors. For each emission, in the left wing it is possible to choose one of the two settings, a_1, a_2, of the left measuring device, and in the right wing—one of the two settings, b_3, b_4, of the right measuring device. Given that the setting selected on the left is a_i, one of the two results, A_i^+ or A_i^-, occurs, and given that the setting selected on the right is b_j, one of the two results, B_j^+ or B_j^- occurs. Some pairs of remote results like A_i^m, B_j^n ($m, n \in \{+, -\}$) are correlated; in the spirit of local causality, we assume "hidden states" (complete states, or, at least, states more complete than the quantum mechanical states), which are supposed to remove the correlations, were the probabilities conditioned on each such a state. By the Bell-Aspect correlations we will understand 16 pairs of the form A_i^m, B_j^n, with their probabilities agreeing with quantum mechanical predictions. Since these 16 pairs can be seen as four groups of correlated pairs connected in the similar way as the pairs featured in Fact 3, by that fact it would be superfluous to consider more than 4 screener systems for them— therefore we posit one screener system for every pair of detector settings.

We model the experiment in a single classical probability space[11] $\langle \Omega, \mathcal{F}, P\rangle$, which of course means that (the representations of) $a_1, a_2, b_3, b_4, A_i^m, B_j^n$ and the hidden states belong to $\mathcal{F}$. These events should satisfy the following natural conditions:

$$A_i^m \subseteq a_i, \qquad B_j^n \subseteq b_j, \qquad a_2 = \Omega \setminus a_1, \qquad b_4 = \Omega \setminus b_3, \quad (5.2)$$

---

[10]From now on, for two events X and Y, $K(XY)$ is a natural number being the size of the screener systems for these two events. We say "the" screener system even though of course many different screener systems may exist for some given events, but one particular will always be intended by the context. If we allow for an infinite screener system, we should understand by $K(XY)$ an index set of cardinality equal to the cardinality of the screener system and write $\alpha \in K(XY)$ rather than $k < K(XY)$. If $X = A_i$ and $Y = B_j$ we will use the expression $K(ij)$.

[11]We choose the single space rather then the many-space approach not because we prefer it (in fact we do not), but because it is employed in the majority of the literature on the subject of the connections between separate- and common common causes (or screener systems) and the Bell inequalities.

$$A_i^- \cup A_i^+ = a_i, \quad B_j^- \cup B_j^+ = b_j \quad \text{for } i = 1, 2; j = 3, 4; m, n \in \{+, -\}. \quad (5.3)$$

Note that this already incorporates some modal claims, e.g., that the result A_i^m must occur in the measurement of a_i (not of a_j), and that A_i^m cannot occur together with A_j^m if $i \neq j$. Notice also that in the single space approach we are using here the Bell-Aspect correlations are *conditional* correlations, for example:

$$P(A_1^+ B_3^+ \mid a_1 b_3) > P(A_1^+ \mid a_1 b_3) P(B_3^+ \mid a_1 b_3),$$

and so we will say that an event C screens off such a correlation between A_1^+ and B_3^+ whenever

$$P(A_1^+ B_3^+ \mid a_1 b_3 C) = P(A_1^+ \mid a_1 b_3 C) P(B_3^+ \mid a_1 b_3 C).$$

Note that due to our just introduced conditions (5.3) and Fact (3), C will also screen off A_1^+ from B_3^-, and so on for other results under the same detector settings.

The conditions PI', OI', and NOCONS are expressed as follows:

$$P(A_i^m \mid a_i b_j C_{ij}^k) = P(A_i^m \mid a_i b_{j'} C_{ij}^k) \qquad \text{PI'}$$

$$P(B_j^n \mid a_i b_j C_{ij}^k) = P(B_j^n \mid a_{i'} b_j C_{ij}^k) \qquad \text{PI'}$$

$$P(A_i^m B_j^n \mid a_i b_j C_{ij}^k) = P(A_i^m \mid a_i b_j C_{ij}^k) P(B_j^n \mid a_i b_j C_{ij}^k) \qquad \text{OI'}$$

$$P(a_i b_j \mathfrak{A}) = P(a_i b_j) P(\mathfrak{A}) \qquad \text{NOCONS}$$

where $\mathfrak{A}$ is any algebraic (Boolean) combination of the elements of four partitions $\{C_{ij}^k\}$ ($i = 1, 2, j = 3, 4$) and all formulas are quantified for all $i, i' \in \{1, 2\}, j, j' \in \{3, 4\}, m, n \in \{+, -\}$ and $k < K(ij)$. We do *not* require a screener system for one correlation to satisfy Parameter Independence (and Outcome Independence) with respect to another correlation, but we briefly explore this option in Sect. 5.3.1. (In Hofer-Szabò's papers, PI' and OI' are called, respectively, Locality and Screening-off.)

Notice that PI' can be equivalently phrased as $P(A_i^m \mid a_i b_j C_{ij}^k) = P(A_i^m \mid a_i C_{ij}^k)$ (similarly for other settings); the setting of the "remote" detector is to be irrelevant for the probability of a given result at the "nearby" detector.

Finally, here are the main definitions of models for Bell-Aspect correlations:

Definition 4 (models with separate or common screener system(s)) Consider a probability space $\langle \Omega, \mathcal{F}, P \rangle$ which contains events a_i (for $i \in \{1, 2\}$) and b_j (for $j \in \{3, 4\}$) corresponding to detector settings and events A_i^m and B_j^n (for i's and j's as before and $m, n \in \{+, -\}$) corresponding to measurement results under the appropriate settings. Suppose that the model exhibits Bell-Aspect correlations.

The probability space is a local non-conspiratorial model with **separate screener systems** (or a **separate-ss model**) for the Bell-Aspect correlations if there exist four partitions of Ω consisting of elements of $\mathcal{F}$, $\{C_{ij}^k\}_{k<K(ij)}$, one for each pair of

detector settings, such that for each such a pair a_i, b_j the partition $\{C_{ij}^k\}_{k < K(ij)}$ meets the conditions of Outcome Independence (OI'), Parameter Independence (PI') and No Conspiracy (NOCONS) with regard to the correlations arising at the detector settings a_i and b_j.

The probability space is a local non-conspiratorial model with **a common screener system** (or a **common-ss model**) for the Bell-Aspect correlations if there exists a single partition of Ω consisting of elements of $\mathcal{F}$, $\{C^k\}$, which meets the conditions of PI', OI', and NOCONS with regard to *all* of those correlations.

We call such models "local" since both OI' and PI' are motivated by locality.

5.3.1 The Different Forms Parameter Independence Can Take

In this subsection we point out that there are alternative ways of formal rendering of the intuition behind the Parameter Independence condition.

As already mentioned, the intuition in question goes roughly as follows: "given *the value(s) of a / the hidden variable(s)* and the measurement setting in one wing of the experiment, the measurement result in that wing is probabilistically independent from the measurement setting in the other wing of the experiment". The different forms of PI' will result from the various ways of spelling out the emphasized fragment of the previous sentence.

Notice that the first line of the above formulation of PI' (p. 92), which we, following other authors, are using in this paper, is equivalent both to

$$P(A_i^m \mid a_i b_j C_{ij}^k) = P(A_i^m \mid a_i C_{ij}^k)$$

and

$$P(A_i^m \mid a_i b_{j'} C_{ij}^k) = P(A_i^m \mid a_i C_{ij}^k) \tag{5.4}$$

(the second line of that definition concerns B_j^n and an analogous discussion holds for it). The original formulation follows from the way of thinking that would have us choose the pair of measurement settings i and j, take the specific hidden variable C_{ij}, and require that given a value of *that*, the probability of a result in one wing does not depend on the measurement setting in the other wing. But (5.4) suggests this might not be the best way to go. Notice that if we take another consequence of PI', plugging "j'" for "j" and vice versa into the formulation on p. 92, obtaining

$$P(A_i^m \mid a_i b_{j'} C_{ij'}^k) = P(A_i^m \mid a_i b_j C_{ij'}^k),$$

we get an equivalent of

$$P(A_i^m \mid a_i b_j C_{ij'}^k) = P(A_i^m \mid a_i C_{ij'}^k) \tag{5.5}$$

Since PI' refers to both settings in the "other" wing, it is not evident why (apart from the measurement settings) only the value of a single hidden variable is conditioned upon. One way to modify this is to refer to any combination of values of the two hidden variables pertaining to the two relevant pairs of measurement settings, obtaining in effect the following notion:

Medium PI': $P(A_i^m \mid a_i b_j C_{ij}^k C_{ij'}^l) = P(A_i^m \mid a_i b_{j'} C_{ij}^k C_{ij'}^l)$, with universal quantification over $i, i' \in \{1, 2\}, j, j' \in \{3, 4\}, m \in \{+, -\}, k < K(ij)$ and $l < K(ij')$. (Similarly for B_j^n.)

However, one could also go "all the way", and say that any combination of the *four* hidden variables should be enough to guarantee that a result in one wing of the experiment is probabilistically independent from the measurement setting in the other wing. That is, one could condition on the same thing as in the requirement of strong no-conspiracy, obtaining the following:

Strong PI': $P(A_i^m \mid a_i b_j \mathfrak{A}) = P(A_i^m \mid a_i b_{j'} \mathfrak{A})$, with universal quantification over $i, i' \in \{1, 2\}, j, j' \in \{3, 4\}, m \in \{+, -\}$, and where $\mathfrak{A}$ is any algebraic combination of the elements of the four partitions $\{C_{ij}^k\}$. (Similarly for B_j^n.)

The feasibility of the various forms of PI' depends on how strongly we think of the hidden variables as capturing aspects of the state of the particle source. Studying the relationship between the various forms of PI' and their connection to other conditions will be a subject of future work. At first glance, they are not equivalent to each other and they seem to be logically independent from both OI' and NOCONS.

We will use the Strong PI' condition in Sect. 5.6.

5.4 A Survey of Earlier Results and the Issue of Reducibility

Probably the first attempt at a local non-conspiratorial separate-ss model for Bell-Aspect correlations was a construction of Szabó (2000). However, it turned out that the model violates NOCONS with respect to intersections of screeners from (separate) screener systems (but satisfies NOCONS with respect to each screener). A significant development was a local non-conspiratorial separate-ss model of Grasshoff et al. (2005) for Bell-Aspect correlations produced in a setup with parallel settings: in this model Bell-Clauser-Horne inequalities are derivable. However, as Hofer-Szabó (2008) showed, Grasshoff et al.'s model is reducible to a model with a common screener system. All that is needed to create a common screener system out of separate screener systems is to take intersections of elements of all separate screener systems, that is, sets like $C_{13}^\alpha \cap C_{23}^\beta \cap C_{14}^\gamma \cap C_{24}^\delta$, i.e., intersections of elements of all the screener systems. Thus, since Grasshoff's model is a common-ss model, the Bell-type inequalities are derivable.

This result casts a new light on the project of constructing separate-ss models for Bell-Aspect correlations: such models should not be reducible to common-ss models. But what does this new desideratum involve? If C_{13}, C_{23}, C_{14}, C_{24} are partitions of the sample space Ω, the set of intersections of their elements, i.e.,

$$C = \{C_{13}^{\alpha} \cap C_{23}^{\beta} \cap C_{14}^{\gamma} \cap C_{24}^{\delta} \mid \alpha < K(13), \beta < K(23), \gamma < K(14), \delta < K(24)\}$$

is also a partition of Ω. Note also that if a separate-ss model satisfies NOCONS with respect to every screener system $\{C_{ij}\}$ ($i = 1, 2, j = 3, 4$), it satisfies NOCONS with respect to the common-ss system C. Thus, the only way that a separate-ss model may stop to reduce to a common-ss model is if PI' or OI' fails with respect to an intersection like the one above. This points to an interesting feature of non-reducible separate-ss models that goes against Bell's intuitions. To recall, Bell believed that the correlation between space-like related events should disappear if a *complete* specification of a state in a relevant region is taken into account (cf. Sect. 5.2). Clearly, an intersection of elements of screener systems corresponds to a more complete state description than a single element of a screener system. Accordingly, non-reducibility entails that less complete states meet PI' and OI', whereas more complete states lose one of these properties. Now, if every local non-conspiratorial separate-ss model for Bell-Aspect correlations was reducible to a common-ss one, that would offer an ironical support for Bell's idea: the most complete state descriptions available in a model with separate screener systems make the correlations disappear. The irony is that the assumptions of the model entail Bell's inequalities.

5.5 Main Results

The overarching goal of the project we are dealing with is to present a *reductio* of the idea of non-conspiratorial separate ss-models for quantum correlations, that is, to show in general that their existence entails the existence of appropriate common ss-models, which are known not to exist. So far we have been unable to achieve that goal in full generality. We present below two theorems relevant to the existence of separate ss-models for quantum correlations. Although, as said, these theorems do not show that the local non-conspiratorial ss-models are not viable, they show to what extend they can be made transformed into the sort of models that entail Bell's inequalities. The first theorem, stated informally, says that if there is a local non-conspiratorial separate-ss model for some Bell-Aspect correlations, then there exists a local (possibly conspiratorial) common screener system model for the same correlations. Our argument will proceed in three steps. Let us explain what roles these steps play.

(1) The algebra $\mathcal{F}$ of a separate-ss model may be arbitrarily large and, in particular, may have no atoms. Our first move is to carve from $\mathcal{F}$ a smaller algebra $\mathcal{F}'$, which has atoms of the form $A_i^m \cap B_j^n \cap C_{13}^{\alpha} \cap C_{23}^{\beta} \cap C_{14}^{\gamma} \cap C_{24}^{\delta}$ and whose every

element is a union of atoms (i.e., $\mathcal{F}'$ is atomistic.[12]) Now, our Lemma 5 says
that if $\langle \Omega, \mathcal{F}, P \rangle$ is a local non-conspiratorial separate-ss model for Bell-Aspect
correlations, then so is $\langle \Omega, \mathcal{F}', P' \rangle$, where $\mathcal{F}'$ is the atomistic algebra carved
from $\mathcal{F}$ and $P' = P_{\mathcal{F}'}$.

(2) $\mathcal{F}'$ may have atoms of probability zero (in such cases P' is called an "unfaithful
measure"). Our next move is to construct a probability space with a faithful
probability measure by removing from $\mathcal{F}'$ all atoms with probability zero; the
new algebra and measure are labelled $\mathcal{F}'^{\mathfrak{F}}$ and $P'^{\mathfrak{F}}$, respectively. Our Lemma 7
then says that if $\langle \Omega', \mathcal{F}', P' \rangle$ is a local non-conspiratorial separate-ss model
for some Bell-Aspect correlations with $\mathcal{F}'$ atomistic, then so is the probability
space $\langle \Omega'^{\mathfrak{F}}, \mathcal{F}'^{\mathfrak{F}}, P'^{\mathfrak{F}} \rangle$ obtained from $\langle \Omega', \mathcal{F}', P' \rangle$, with the faithful measure $P'^{\mathfrak{F}}$
assigning to every "new" event the measure of its "original version" given by P.

(3) In the final step we transform the space we constructed in the second step
into a local common-ss model for the correlations. Our Theorem 1 says that
if $\langle \Omega, \mathcal{F}, P \rangle$ is a local non-conspiratorial separate-ss model for Bell-Aspect
correlations with $\mathcal{F}$ atomistic and having atoms of a specific form, while P is a
faithful measure on $\mathcal{F}$, then there is a local (possibly conspiratorial) common-ss
model $\langle \Omega^*, \mathcal{F}^*, P^* \rangle$ for the same Bell-Aspect correlations.

Our second theorem states that any model with separate screener systems for
the Bell-Aspect correlations that satisfy strong PI and strong NOCONS can be
transformed into a model for the same correlations, in which PI is somewhat
compromised, strong NOCONS holds, and the elements of the constructed partition
act deterministically with respect to measurement results.

Since screener systems investigated in the literature are typically finite, our
proofs assume a finite number of screeners in each screener system. However,
our theorems remain correct for infinitely large screener systems (we have already
remarked that they can be at most countably infinite). In some footnotes we indicate
how to modify our proofs for the general case.

5.5.1   To an Atomistic Algebra of Events

The algebra $\mathcal{F}$ of a separate-ss model $\langle \Omega, \mathcal{F}, P \rangle$ for Bell-Aspect correlations might
be very large; without any loss of generality, we may consider its "pruned" cousin, in
which a (new) algebra $\mathcal{F}'$ will be atomistic with atoms being the nonempty elements
of the following set[13]:

---

[12] Recall: an algebra of sets is *atomic* if for every non-minimal element p there exists an atom a
such that $a \subseteq p$; an algebra of sets is *atomistic* if it is atomic and such that every non-minimal
element is a union of atoms.

[13] If we allow for infinite screener systems, we should write $\alpha \in K(13)$, etc., as suggested in
Footnote 10.

$$\mathcal{A} := \{A_i^m \cap B_j^n \cap C_{13}^\alpha \cap C_{14}^\beta \cap C_{23}^\gamma \cap C_{24}^\delta \mid m, n \in \{+, -\}, i \in \{1, 3\},$$
$$j \in \{2, 4\}, \alpha < K(13), \beta < K(14), \gamma < K(23) \text{ and } \delta < K(24)\} \tag{5.6}$$

Lemma 5 *Let $\langle \Omega, \mathcal{F}, P \rangle$ be a local non-conspiratorial separate-ss model for Bell-Aspect correlations. Then the probability space $\langle \Omega, \mathcal{F}', P' \rangle$ also is a local non-conspiratorial separate-ss model for (the same) Bell-Aspect correlations, where*

- $X \in \mathcal{F}'$ *iff for some $A \subseteq \mathcal{A}$ of Eq. 5.6:* $\cup A = X$;
- $P' = P_{|\mathcal{F}'}$.

Proof Immediate. Notice that the first item above says in effect that $\mathcal{F}'$ is atomistic.

5.5.2   From an Unfaithful to Faithful Measure

The measure P of a local non-conspiratorial separate-ss model $\langle \Omega, \mathcal{F}, P \rangle$ need not be faithful, that is, it may be that for some nonempty $X \in \mathcal{F}: P(X) = 0$. Our final construction, however, requires probability spaces with faithful measures. The algebra of events of the model whose existence is guaranteed by Lemma 5 is atomistic and has countably many atoms (since all four screener systems involved are countable). In such a case there is a simple procedure of arriving at another probability space which will also be a local non-conspiratorial separate-ss model for the same correlations, but whose measure will be faithful.

Suppose $\langle \Omega, \mathcal{F}, P \rangle$ is a probability space, with $\mathcal{F}$ atomistic and having countably many atoms. Let $\mathcal{A}$ be the set of atoms of $\mathcal{F}$, and let $\mathcal{A}^+$ be the set of atoms of $\mathcal{F}$ whose probability is greater than 0. Since $\mathcal{A}$ is countable, $\mathcal{A}^+$ is not empty. Atomicity means that for any $E \in \mathcal{F}$ there exists exactly one set $\mathcal{A}_E \subseteq \mathcal{A}$ such that $E = \cup \mathcal{A}_E$. Consider a function f with domain $\mathcal{F}$ defined in the following way: for $E \in \mathcal{F}, f(E) = \cup(\mathcal{A}_E \cap \mathcal{A}^+)$. In effect, the function f "strips down" events (which are unions of atoms, due to $\mathcal{F}$ being atomistic) of their zero-measure parts. The algebra of events of the new space, $\mathcal{F}^{\mathfrak{F}}$, will be simply the image of $\mathcal{F}$ through the function $f: \mathcal{F}^{\mathfrak{F}} = f(\mathcal{F})$. The probability function $P^{\mathfrak{F}}$ assigns to all events $f(E)$ the measure of E in the original space; the important difference is that if $E \in \mathcal{F}$ is a nonempty measure zero event, then $f(E) = \emptyset$, which ensures that $P^{\mathfrak{F}}$ is faithful.

Definition 6 (faithfulisation)   Let $\mathcal{S} = \langle \Omega, \mathcal{F}, P \rangle$ be a probability space with $\mathcal{F}$ atomistic and having countably many atoms. Let $\mathcal{A}$ be the set of atoms of $\mathcal{F}$, and let $\mathcal{A}^+$ be the set of atoms of $\mathcal{F}$ whose probability is greater than 0. For any $E \in \mathcal{F}$, let $\mathcal{A}_E$ be the subset of $\mathcal{A}$ such that $E = \cup \mathcal{A}_E$. Consider a function $f : \mathcal{F} \to \Omega$: $f(E) = \cup(\mathcal{A}_E \cap \mathcal{A}^+)$.

The faithfulisation of $\mathcal{S}$ is a triple $\langle \Omega^{\mathfrak{F}}, \mathcal{F}^{\mathfrak{F}}, P^{\mathfrak{F}} \rangle$, where:

- $\mathcal{F}^{\mathfrak{F}} = f(\mathcal{F})$;
- $P^{\mathfrak{F}}(f(E)) = P(E)$;
- $\Omega^{\mathfrak{F}} = f(\Omega)$.

A simple proof of the following lemma is left to the Reader:

Lemma 7 *If $\langle \Omega, \mathcal{F}, P \rangle$ is a probability space such that $\mathcal{F}$ is atomistic and has countably many atoms, then its faithfulisation $\langle \Omega^{\mathfrak{F}}, \mathcal{F}^{\mathfrak{F}}, P^{\mathfrak{F}} \rangle$ is a probability space as well. Moreover, $P^{\mathfrak{F}}$ is faithful. If $\langle \Omega, \mathcal{F}, P \rangle$ is a local, non-conspiratorial separate-ss model for Bell-Aspect correlations, then its faithfulisation also is a local, non-conspiratorial separate-ss model for the same correlations.*

5.5.3 From Separate-ss Models to Common-ss Models

Theorem 1 *Let $(\Omega, \mathcal{F}, P)$ be a local non-conspiratorial separate-ss model for Bell-Aspect correlations with faithful measure P and $\mathcal{F}$ atomistic and having the countable set $\mathcal{A}'$ of atoms such that*

$$\mathcal{A}' \subseteq \{A_i^m \cap B_j^n \cap C_{1,3}^\alpha \cap C_{1,4}^\beta \cap C_{2,3}^\gamma \cap C_{2,4}^\delta \mid m, n \in \{+, -\}, i \in \{1, 3\},$$

$$j \in \{2, 4\}, \alpha \leq K(1,3), \beta \leq K(1,4), \gamma \leq K(2,3) \text{ and } \delta \leq K(2,4)\}.$$

Then there is a local common-ss model $(\Omega', \mathcal{F}', P')$ for the same Bell-Aspect correlations with P' being faithful and such that there exists an embedding $\varphi : \mathcal{F} \to \mathcal{F}'$ such that

$$P(A_i^m \cap B_j^n \cap a_i \cap b_j) = P'(\varphi(A_i^m \cap B_j^n \cap a_i \cap b_j)).$$

Proof Let us first define the following four functions:

$$\wedge (A_i^m, B_j^n) := A_i^m \cap B_j^n, \quad L(A_i^m, B_j^n) := A_i^m \cap (b_j \setminus B_j^n),$$

$$R(A_i^m, B_j^n) := (a_i \setminus A_i^m) \cap B_j^n, \text{ and } \emptyset(A_i^m, B_j^n) := (a_i \setminus A_i^m) \cap (b_j \setminus B_j^n)$$

where $i \in \{1, 2\}; j \in \{3, 4\}; m, n \in \{+, -\}$.

It will also be convenient to gather the names of the four functions (we will omit any quoting devices) into the set $I := \{\wedge, L, R, \emptyset\}$.

The task is now to construct the probability space $\langle \Omega', \mathcal{F}', P' \rangle$. Let Ω' be an infinite set. $\mathcal{F}'$ will be an atomistic countable algebra of subsets of Ω'. We will refer to the atoms of $\mathcal{F}'$ by labels of the following sort:

$$_{ij}a_{xyzt}^{\alpha\beta\gamma\delta}, \text{ where } i \in \{1, 2\}, j \in \{3, 4\}, \alpha < K(13), \beta < K(14),$$

$$\gamma < K(23), \delta < K(24), x, y, z, t \in I.$$

However, we will consider screener systems as a particular sort of instruction sets and thus, admit only certain sets (referred to by respective labels) as the atoms of the constructed algebra—namely the ones that are coherent in the view of screener systems as instruction sets.

First, for $f \in I$ let us denote by $\pi_A(f(A_i^m, B_j^n))$ and $\pi_B(f(A_i^m, B_j^n))$ the A-th and respectively the B-th component of the image of (A_i^m, B_j^n) under the function f, e.g. $\pi_A(R(A_1^+, B_3^-)) = a_1 \setminus A_1^+$ and $\pi_B(\wedge(A_2^+, B_4^-)) = B_4^-$.

We will say that the label a_{xyzt} is *coherent* if it satisfies the following conditions:

$$\pi_A(x(A_1^+, B_3^+)) = \pi_A(y(A_1^+, B_4^+)),$$
$$\pi_A(z(A_2^+, B_3^+)) = \pi_A(t(A_2^+, B_4^+)),$$
$$\pi_B(x(A_1^+, B_3^+)) = \pi_B(z(A_2^+, B_3^+)),$$
$$\pi_B(y(A_1^+, B_4^+)) = \pi_B(z(A_2^+, B_4^+)).$$

Thus, to consider an example, let us see what are the coherent labels y, z, t for the *potential* atom $a_{\wedge yzt}$:

$$x(A_1^+, B_3^+) = \wedge(A_1^+, B_3^+),$$

therefore, since $\pi_A(\wedge(A_1^+, B_3^+)) = A_1^+$, the only admissible labels for y are the following:

$$y(A_1^+, B_4^+) = \wedge(A_1^+, B_4^+)$$

or

$$y(A_1^+, B_4^+) = L(A_1^+, B_4^+).$$

If the first labeling is chosen, then we have $a_{\wedge\wedge zt}$ and since $\pi_B(\wedge(A_1^+, B_3^+)) = B_3^+$, the value of z may be $\wedge$ or R, because otherwise $\pi_B(\wedge(A_2^+, B_3^+)) = b_3 \setminus B_3^+$ which would result in the label being incoherent. If $y = \wedge$ and $z = \wedge$, then $t = \wedge$. If $y = \wedge$ and $z = R$, then $t = R$. On the other hand, if $y = L$, then the possible values of z are R and $\wedge$. Then, the value of t would be respectively $\emptyset$ or L.

Applying the coherence condition to the labels of the potential atoms, we obtain the possible labels of the atom $a_{\wedge yzt}$:

$$a_{\wedge\wedge\wedge\wedge} \quad (A_1^+, B_3^+, A_2^+, B_4^+),$$
$$a_{\wedge\wedge RR} \quad (A_1^+, B_3^+, A_2^-, B_4^+),$$
$$a_{\wedge LR\emptyset} \quad (A_1^+, B_3^+, A_2^-, B_4^-),$$
$$a_{\wedge L\wedge L} \quad (A_1^+, B_3^+, A_2^+, B_4^-).$$

Observe also that not all labels of this sort will denote atoms of $\mathcal{F}'$. The sole class of exceptions are the labels $_{ij}a_{xyzt}^{\alpha\beta\gamma\delta}$ for which $A_i^m \cap B_j^n \cap C_{13}^\alpha \cap C_{14}^\beta \cap C_{23}^\gamma \cap C_{24}^\delta \notin \mathcal{A}'$. (In other words, $A_i^m \cap B_j^n \cap C_{13}^\alpha \cap C_{14}^\beta \cap C_{23}^\gamma \cap C_{24}^\delta = \emptyset \in \mathcal{F}$.)[14] Such labels will refer to $\emptyset \in \mathcal{F}'$.

The measure P' is determined by assigning the following measure to the atoms of $\mathcal{F}'$:

$$P'(_{ij}a_{xyzt}^{\alpha\beta\gamma\delta}) = \frac{1}{4} \cdot P(a_i b_j \cap f(A_i B_j) \cap C_{1,3}^\alpha \cap C_{1,4}^\beta \cap C_{2,3}^\gamma \cap C_{2,4}^\delta),$$

where f_{ij} is a function from I appropriate for the pair ij, i.e. $f_{13} = x, f_{14} = y, f_{23} = z$ and $f_{24} = t$. The factor $\frac{1}{4}$ is chosen in such a way that values of P' are in a sense *symmetric* with respect to P while computing them for various pairs of measurement settings ij.

Since $\mathcal{F}'$ is atomistic, the measure on all its elements is determined.

It is easy to check that P' is faithful.

The embedding $\varphi : \mathcal{F} \to \mathcal{F}'$ is determined by its action on the atoms of $\mathcal{F}$:

$$\text{for } f_{ij} \in I, \text{ if } C_{i,j}^\alpha \cap C_{1,4}^\beta \cap C_{2,3}^\gamma \cap C_{2,4}^\delta \cap f_{ij}(A_i^+, B_j^+) \neq \emptyset, \text{ then}$$

$$\varphi(C_{1,3}^\alpha \cap C_{1,4}^\beta \cap C_{2,3}^\gamma \cap C_{2,4}^\delta \cap f_{ij}(A_i^+, B_j^+)) = \bigcup_{I \setminus \{f_{ij}\}} {}_{ij}a_{xyzt}^{\alpha\beta\gamma\delta}.$$

with f_{ij} fixed in all the summed a_{xyzt} and all the sums defined exclusively for atoms with coherent labels.

For example:

$$\text{for any } x \in I, \text{ if } C_{1,3}^\alpha \cap C_{1,4}^\beta \cap C_{2,3}^\gamma \cap C_{2,4}^\delta \cap x(A_1^+, B_3^+) \neq \emptyset, \text{ then}$$

$$\varphi(C_{1,3}^\alpha \cap C_{1,4}^\beta \cap C_{2,3}^\gamma \cap C_{2,4}^\delta \cap x(A_1^+, B_3^+)) = \bigcup_{y,z,t \in I} {}_{13}a_{xyzt}^{\alpha\beta\gamma\delta};$$

Next, by the definition of P' (here dividing P by 4 turns out useful), the measure P on elements $A_i^m \cap B_j^n \cap a_i \cap b_j \in \mathcal{F}$ agrees with the measure P' on the φ-images of these elements, since the latter are unions of atoms with coherent labels and so P agrees with P' also on the images of elements of the form $a_i \cap b_j$.

We now introduce the common screener system. The set S below is a common screener system for the correlations and each element of S satisfies OI' and PI':

---

[14]This may happen e.g. when the correlation between A_i^m and B_j^n is perfect, i.e., when $P(A_i^m B_j^n \mid a_i b_j) = 1$. Even though such a case is experimentally unrealisable, we cater to it for more generality.

$$S := \{S^{\alpha\beta\gamma\delta}_{xyzt} \mid \alpha \leqslant K(1,3), \beta \leqslant K(1,4), \gamma \leqslant K(2,3), \delta \leqslant K(2,4), x,y,z,t \in I\}$$
$$\text{where } S^{\alpha\beta\gamma\delta}_{xyzt} := {}_{13}a^{\alpha\beta\gamma\delta}_{xyzt} \cup {}_{14}a^{\alpha\beta\gamma\delta}_{xyzt} \cup {}_{23}a^{\alpha\beta\gamma\delta}_{xyzt} \cup {}_{24}a^{\alpha\beta\gamma\delta}_{xyzt} .$$

Indeed: S is a partition and OI' and PI' are actually algebraic conditions guaranteed by the consistent labeling. Note however that NOCONS may fail. **Q.E.D.**

5.6 Strong PI and Partially Deterministic Separate-ss Models

We will now show a way of transforming a model with separate screener systems for the Bell-Aspect correlations which satisfy Strong PI' (introduced in Sect. 5.3.1) and NOCONS into a somewhat different model in which strong PI' is somewhat compromised, but NOCONS and "regular" PI' hold, and the elements of the constructed partition act deterministically with respect to the measurement results. The idea is to continue the line of research started by Hofer-Szabó (2012), in particular in his Proposition 1 (p. 120), which deals with deterministic screener systems.

This section is devoted to the proof of the following theorem:

Theorem 1 *If a model for Bell-Aspect correlations with four partitions corresponding to four pairs of measurement settings satisfies strong PI' and NOCONS, then it allows the derivation of the CHSH inequalities.*

In the initial model the role of "separate screener systems" is to be played by four finite partitions $C_{ij} = \{C^{\sigma}_{ij}\}_{\sigma < K(ij)}$; $K(ij)$ is the cardinality of the partition for the pair of detector settings a_i and b_j. Regarding the formulas immediately below this paragraph, assume that $\mathfrak{A}$ is any Boolean combination of the elements of the four partitions C_{ij}, and that all formulas are quantified for all $i, i' \in \{1,2\}, j, j' \in \{3,4\}$, $m, n \in \{+,-\}, \sigma < K(ij)$, and all $\mathfrak{A}$ (in some cases the quantification will be vacuous). The conditions of PI' and NOCONS in strong and weak versions (note: "weak PI' " is just the original PI' from p. 92) are expressed as follows:

$$P(A^m_i \mid a_ib_jC^{\sigma}_{ij}) = P(A^m_i \mid a_ib_{j'}C^{\sigma}_{ij}) \qquad \text{(weak PI', part I)}$$
$$P(B^n_j \mid a_ib_jC^{\sigma}_{ij}) = P(B^n_j \mid a_{i'}b_jC^{\sigma}_{ij}) \qquad \text{(weak PI', part II)}$$
$$P(A^m_i \mid a_ib_j\mathfrak{A}) = P(A^m_i \mid a_ib_{j'}\mathfrak{A}) \qquad \text{(strong PI', part I)}$$
$$P(B^n_j \mid a_ib_j\mathfrak{A}) = P(B^n_j \mid a_{i'}b_j\mathfrak{A}) \qquad \text{(strong PI', part II)}$$
$$P(a_ib_jC^{\sigma}_{ij}) = P(a_ib_j)P(C^{\sigma}_{ij}) \qquad \text{(weak NOCONS)}$$
$$P(a_ib_j\mathfrak{A}) = P(a_ib_j)P(\mathfrak{A}) \qquad \text{(NOCONS).}$$

We assume that the probability space we are dealing with is atomless, in the sense that for any event E such that $P(E) > 0$ and for any $0 < x < P(E)$ there exists an event F such that $F \subset E$ and $P(F) = x$. This does not result in a loss of generality since any probability space can be embedded in one atomless in the above sense.

The construction. We first focus on the partition $\{C_{13}^\alpha\}$. For notational clarity we assume these abbreviations: $A_1 B_3 = A_1^+ B_3^+$, $C^\alpha = C_{13}^\alpha$ and $C^{\beta\gamma\delta} := C_{14}^\beta C_{23}^\gamma C_{24}^\delta$.

We pick some $\alpha, \beta, \gamma, \delta$ from their respective ranges and postulate a partition $\{C^{\alpha\wedge}, C^{\alpha L}, C^{\alpha R}, C^{\alpha\emptyset}\}$ of C^α which, provided $p(a_1 b_3 C^\alpha C^{\beta\gamma\delta}) > 0$, satisfies for every m, n:

$$p(y(A_1 B_3) \mid a_1 b_3 C^{\alpha x} C^{\beta\gamma\delta}) = \begin{cases} 1 \text{ if } x = y \\ 0 \text{ otherwise.} \end{cases} \tag{5.7}$$

We assume that (†) $_{13}C^{\alpha x \beta\gamma\delta} := x(A_1 B_3) C^\alpha C^{\beta\gamma\delta}$ is a part of $C^{\alpha x}$ contained in the intersection $a_1 b_3 C^\alpha C^{\beta\gamma\delta}$. (If $p(a_1 b_3 C^\alpha C^{\beta\gamma\delta}) = 0$, we set $_{13}C^{\alpha x \beta\gamma\delta} := \emptyset$). In what follows, we will construct sets $_{ij}C^{\alpha x\beta\gamma\delta}$, that is, parts of $C^{\alpha x}$ contained in the remaining three events of the form $a_i b_j C^\alpha C^{\beta\gamma\delta}$. The whole $C^{\alpha x}$ then is:

$$C^{\alpha x} = \bigcup_{ij\beta\gamma\delta} {}_{ij}C^{\alpha x\beta\gamma\delta} \tag{5.8}$$

By strong PI' and NOCONS,

$$p(A_1^+ a_1 b_3 C^\alpha C^{\beta\gamma\delta}) p(a_1 b_4) = p(A_1^+ a_1 b_4 C^\alpha C^{\beta\gamma\delta}) p(a_1 b_3)$$

By (†) $(A_1^+ a_1 b_3 C^\alpha C^{\beta\gamma\delta}) = {}_{13}C^{\alpha L\beta\gamma\delta} \cup {}_{13}C^{\alpha\wedge\beta\gamma\delta}$, so in virtue of strong PI' and strong NOCONS,

$$p(A_1^+ a_1 b_4 C^\alpha C^{\beta\gamma\delta}) = (p({}_{13}C^{\alpha L\beta\gamma\delta}) + p({}_{13}C^{\alpha\wedge\beta\gamma\delta})) p(a_1 b_4)/p(a_1 b_3).$$

We next partition $A_1^+ a_1 b_4 C^\alpha C^{\beta\gamma\delta}$ into $_{14}C^{\alpha L\beta\gamma\delta}$ and $_{14}C^{\alpha\wedge\beta\gamma\delta}$ such that (since the initial probability space is non-atomic)

$$p({}_{14}C^{\alpha L\beta\gamma\delta}) = p({}_{13}C^{\alpha L\beta\gamma\delta}) p(a_1 b_4)/p(a_1 b_3) \tag{5.9}$$

$$p({}_{14}C^{\alpha\wedge\beta\gamma\delta}) = p({}_{13}C^{\alpha\wedge\beta\gamma\delta}) p(a_1 b_4)/p(a_1 b_3). \tag{5.10}$$

By working out the A_1^- case, we get

$$A_1^- a_1 b_4 C^\alpha C^{\beta\gamma\delta} = {}_{14}C^{\alpha R\beta\gamma\delta} \cup {}_{14}C^{\alpha\emptyset\beta\gamma\delta} \tag{5.11}$$

$$p({}_{14}C^{\alpha R\beta\gamma\delta}) = p({}_{13}C^{\alpha R\beta\gamma\delta}) p(a_1 b_4)/p(a_1 b_3) \tag{5.12}$$

$$p({}_{14}C^{\alpha\emptyset\beta\gamma\delta}) = p({}_{13}C^{\alpha\emptyset\beta\gamma\delta}) p(a_1 b_4)/p(a_1 b_3). \tag{5.13}$$

In a similar vein we analyze the $(2, 3)$ pair of settings, obtaining:

$$B_3^+ a_2 b_3 C^\alpha C^{\beta\gamma\delta} = {}_{23}C^{\alpha R\beta\gamma\delta} \cup {}_{23}C^{\alpha\wedge\beta\gamma\delta} \tag{5.14}$$

$$p({}_{23}C^{\alpha R\beta\gamma\delta}) = p({}_{13}C^{\alpha R\beta\gamma\delta})p(a_2 b_3)/p(a_1 b_3) \tag{5.15}$$

$$p({}_{23}C^{\alpha\emptyset\beta\gamma\delta}) = p({}_{13}C^{\alpha\emptyset\beta\gamma\delta})p(a_2 b_3)/p(a_1 b_3) \tag{5.16}$$

$$B_3^- a_2 b_3 C^\alpha C^{\beta\gamma\delta} = {}_{23}C^{\alpha L\beta\gamma\delta} \cup {}_{23}C^{\alpha\emptyset\beta\gamma\delta} \tag{5.17}$$

$$p({}_{23}C^{\alpha L\beta\gamma\delta}) = p({}_{13}C^{\alpha L\beta\gamma\delta})p(a_2 b_3)/p(a_1 b_3) \tag{5.18}$$

$$p({}_{23}C^{\alpha\emptyset\beta\gamma\delta}) = p({}_{13}C^{\alpha\emptyset\beta\gamma\delta})p(a_2 b_3)/p(a_1 b_3). \tag{5.19}$$

As for the pair of settings $(2, 4)$, by virtue of NOCONS, i.e,

$$p(C^\alpha C^{\beta\gamma\delta}) = p(a_2 b_4 C^\alpha C^{\beta\gamma\delta})/p(a_2 b_4) = p(a_1 b_3 C^\alpha C^{\beta\gamma\delta})/p(a_1 b_3),$$

we divide $a_2 b_4 C^\alpha C^{\beta\gamma\delta}$ proportionally into ${}_{24}C^{\alpha x\beta\gamma\delta}$, i.e.,:

$$a_2 b_4 C^\alpha C^{\beta\gamma\delta} = \bigcup_x {}_{24}C^{\alpha x\beta\gamma\delta} \tag{5.20}$$

$$p({}_{24}C^{\alpha x\beta\gamma\delta}) = p({}_{13}C^{\alpha x\beta\gamma\delta})p(a_2 b_4)/p(a_1 b_3) \tag{5.21}$$

Having constructed parts of $C^{\alpha x}$ in all $a_i b_j C^\alpha C^{\beta\gamma\delta}$, we obtain $C^{\alpha x}$ by Eq. 5.8. We repeat this construction for each $\alpha \in K(13)$, the result being these four partitions:

$$\{C_{13}^{\alpha x}\}_{\alpha \in K(13), x \in I} \quad \{C_{14}^\beta\}_{\beta \in K(14)} \quad \{C_{23}^\gamma\}_{\gamma \in K(23)} \quad \{C_{24}^\delta\}_{\delta \in K(24)}. \tag{5.22}$$

We next check that the above system of partitions satisfies what we promised it would satisfy.

First, Eq. 5.7 guarantees that each $C^{\alpha x} := C_{13}^{\alpha x}$ acts deterministically with respect to the measurement results, that is, that,

$$p(y(A_1 B_3) \mid a_1 b_3 C_{13}^{\alpha x}) = \begin{cases} 1 \text{ if } x = y \\ 0 \text{ otherwise.} \end{cases} \tag{5.23}$$

Second, to show that NOCONS holds, we calculate

$$p(C^{\alpha x} C^{\beta\gamma\delta}) = \sum_{ij} p({}_{ij}C^{\alpha x\beta\gamma\delta}) =$$

$$p({}_{13}C^{\alpha x\beta\gamma\delta})(p(a_1 b_3) + p(a_1 b_4) + p(a_2 b_3) + p(a_2 b_4))/p(a_1 b_3) =$$

$$p({}_{13}C^{\alpha x\beta\gamma\delta})/p(a_1 b_3) = p(C^{\alpha x} C^{\beta\gamma\delta} a_1 b_3)/p(a_1 b_3).$$

$$\tag{5.24}$$

By Eqs. 5.3–5.14 we similarly get

$$p(C^{\alpha x} C^{\beta \gamma \delta}) = p(_{14}C^{\alpha x \beta \gamma \delta})/p(a_1 b_4) = p(C^{\alpha x} C^{\beta \gamma \delta} a_1 b_4)/p(a_1 b_4) \tag{5.25}$$

$$p(C^{\alpha x} C^{\beta \gamma \delta}) = p(_{23}C^{\alpha x \beta \gamma \delta})/p(a_2 b_3) = p(C^{\alpha x} C^{\beta \gamma \delta} a_2 b_3)/p(a_2 b_3) \tag{5.26}$$

$$p(C^{\alpha x} C^{\beta \gamma \delta}) = p(_{24}C^{\alpha x \beta \gamma \delta})/p(a_2 b_4) = p(C^{\alpha x} C^{\beta \gamma \delta} a_2 b_4)/p(a_2 b_4). \tag{5.27}$$

Finally, we see that the partitions satisfy a restricted version of strong PI', that is, one restricted to the Boolean combinations of values of the screener systems for the pairs $(1, 3)$ and $(1, 4)$ and also for the pairs $(1, 3)$ and $(2, 3)$. By Eq. 5.7 and in virtue of our construction of $_{23}C^{\alpha x \beta \gamma \delta}$ and $_{14}C^{\alpha x \beta \gamma \delta}$ we have, for every $m, n, x, \alpha, \beta, \gamma, \delta$:

$$p(A_1^m \mid a_1 b_3 C^{\alpha x} C^{\beta \gamma \delta}) = p(A_1^m \mid a_1 b_4 C^{\alpha x} C^{\beta \gamma \delta}) = 1 \text{ or } 0; \tag{5.28}$$

$$p(B_3^n \mid a_1 b_3 C^{\alpha x} C^{\beta \gamma \delta}) = p(B_3^n \mid a_2 b_3 C^{\alpha x} C^{\beta \gamma \delta}) = 1 \text{ or } 0. \tag{5.29}$$

However, strong PI can fail for combinations of values of the screener systems for the pairs $(1, 4)$ and $(2, 4)$, and also for the pair $(2, 3)$ and $(2, 4)$. For instance, $p(A_2^+ \mid a_2 b_3 C^{1L} C^{\beta \gamma \delta}) = 1$, whereas $p(A_2^+ \mid a_2 b_4 C^{1L} C^{\beta \gamma \delta}) = p(A_2^+ \, _{24}C^{1L \beta \gamma \delta})/p(\, _{24}C^{1L \beta \gamma \delta})$ can be any probability.

However, weak PI is satisfied, as the following Lemma shows:

Lemma 8 *If a separate ss model for Bell-Aspect correlations satisfies strong PI', OI', and NOCONS, then its extension described above satisfies weak PI'.*

Proof Observe that weak PI' relates pairs of measurement settings with one setting in common and involves conditioning on an element of an "old" partition (that is, of $\{C_{14}^\beta\}$ or $\{C_{23}^\gamma\}$ or $\{C_{24}^\delta\}$) or an element of the "new" partition $\{C_{13}^{\alpha x}\}_{\alpha \in K(1,3), x \in I}$. If the conditioning is on an element of the old partition, weak PI holds, because in the initial model strong PI, and hence weak PI are satisfied. Thus, the formulas like the following are true in the extended model:

$$p(A_2^m \mid a_2 b_3 C_{23}^\gamma) = p(A_2^m \mid a_2 b_4 C_{23}^\gamma) \tag{5.30}$$

$$p(A_1^m \mid a_1 b_4 C_{14}^\gamma) = p(A_1^m \mid a_1 b_3 C_{14}^\gamma). \tag{5.31}$$

Turning next to cases with conditioning on an element of the "new" partition, we consider the cases relating the measurement settings (a_1, b_3) and (a_1, b_4), and so restricting our attention to the left result, i.e., A_1^m. The remaining cases are calculated analogously. Thus, our task is to check if this formula holds in the extended model:

$$p(A_1^m \mid a_1 b_3 C_{13}^{\alpha x}) = p(A_1^m \mid a_1 b_4 C_{13}^{\alpha x}). \tag{5.32}$$

We first introduce a piece of notation: we say that $m \in \{-1, +1\}$ and $x \in I$ agree, $\mathfrak{a}(m, x)$, iff $m = +1$ and $x \in \{\wedge, L\}$ or $m = -1$ and $x \in \{R, \emptyset\}$. Clearly, by the construction of $\{C_{13}^{\alpha x}\}$, (i) LHS of Eq. 5.32 is 1 if $\mathfrak{a}(m, x)$; otherwise it is 0.

Turning to the RHS of Eq. 5.32, note that $a_1 b_4 C_{13}^{\alpha x} = \bigcup_{\beta\gamma\delta} a_1 b_4 C_{13}^{\alpha x \beta \gamma \delta}$. And $A_1^m a_1 b_4 C_{13}^{\alpha x} = \bigcup_{\beta\gamma\delta} a_1 b_4 C_{13}^{\alpha x \beta \gamma \delta}$ if $\mathfrak{a}(m,x)$. Otherwise $A_1^m a_1 b_4 C_{13}^{\alpha x} = \emptyset$. Hence, (ii) the RHS of Eq. 5.32 is 1 if $\mathfrak{a}(m,x)$; otherwise it is 0. Comparing (i) and (ii) we see that Eq. 5.32 holds. **Q.E.D.**

To return to the construction, we need to produce a second system of partitions, satisfying analogous conditions. The aim is to get four partitions (we will now fine-grain the initial partition $\{C_{24}^{\delta}\}_{\delta \in K(24), x \in \Lambda}$):

$$\{C_{13}^{\alpha}\}_{\alpha \in K(13)} \quad \{C_{14}^{\beta}\}_{\beta \in K(14)} \quad \{C_{23}^{\gamma}\}_{\gamma \in K(23)} \quad \{C_{24}^{\delta x}\}_{\delta \in K(24), x \in \Lambda}.$$

To this end, working on the algebra generated by elements of the four partitions from Eq. (5.22), settings, and results, we do the same construction with respect to the partitions $\{C_{13}^{\alpha}\}_{\alpha \in K(13)}$, $\{C_{14}^{\beta}\}_{\alpha \in K(14)}$, $\{C_{23}^{\gamma}\}_{\alpha \in K(23)}$, and $\{C_{24}^{\delta}\}_{\alpha \in K(24)}$, fine-graining the last one. Note that since our goal is to connect this with Proposition 1 from Hofer-Szabó (2012), we are only doing the construction for two pairs of measurement settings, and not for all four, since this is enough to proceed with the proof of that Proposition. Our construction results in a separate-ss model which satisfies weak PI', NOCONS, and such that all elements of two of the separate screener systems, namely those for pairs (a_1, b_3) and (a_2, b_4), act deterministically on the measurement results. This means that the model satisfies the actual assumptions of Hofer-Szabó's Proposition 1, and so, allows the derivation of the Clauser-Horne-Shimony-Holt (CHSH) inequalities. Since these are falsifiable for some experimental setups, then for those setups *no models for Bell-Aspect correlations which would satisfy strong PI' and NOCONS exist.* As advertised at the beginning of this section, to sum up, we have shown the following theorem:

Theorem 2 *If a model for Bell-Aspect correlations with four partitions corresponding to four pairs of measurement settings satisfies strong PI' and NOCONS, then it allows the derivation of the CHSH inequalities.*

Notice that in the effort to be as general as possible, in this theorem we do not call the initial model a "separate-ss model", because we do not need to assume that the elements of the initial four partitions satisfy OI'.

5.7 Conclusions

The essence of the project we have analysed is the distinction between (many) separate screener systems and (single) common screener system. Importantly, the condition of Outcome Independence present in the usual derivations of Bell's theorem pertains to a common screener system since it requires that a posited set of hidden states forms a common screener system for all correlated pairs of results. This brings in two separate questions. First, are models with several separate screener systems for the Bell-Aspect experiment more general than models with a common screener system for the same experiment? And, secondly, does the distinction between separate screener systems and common screener system explain why the Bell inequalities are violated? The explanation we are alluding to should argue that OI (which is needed to derive the Bell inequalities) is not reasonable for some reasons and its substitute, OI', is reasonable, but does not lead to the inequalities.

In relation to the first question, our proofs offer detailed insights into the relations between two sorts of models for the experiment. The first one establishes, that if for some Bell-Aspect correlations there is a local non-conspiratorial model with separate screener systems, then there is a local model with common screener system for the same correlations, though in that resulting model NOCONS can be compromised. This means that in these models Bell's inequalities might happen not to be derivable. Thus, strictly speaking, we failed to show that any model with several separate screener systems is reducible to a model with a common screener system that satisfies the premises of Bell's inequalities. Neither have we found evidence to the contrary, i.e., that the former models are not reducible to the latter. The question of the reducibility remains thus an open problem for future research. Still, we managed to show another class of models with different hidden variables corresponding to different pairs of measurement settings in which the CHSH inequalities are derivable: strong PI' and NOCONS are enough for this.

As for the second question, our paper adds a little to the existing results of Grasshoff et al. (2005) and Hofer-Szabó (2011). To recall, the first paper shows that a model with several separate screener systems for Bell-Aspect experiment with parallel settings reduces to a model with a common screener system, which in turn permits a derivation of a Bell inequality. The assumption of parallel settings is relaxed in the second paper, which shows that Bell inequalities are still derivable for non-parallel settings, provided that the settings are sufficiently close to being parallel. Thus, in some circumstances OI' suffices to derive the Bell inequalities, so it can hardly explain why these inequalities are violated.

References

Aspect, A., J. Dalibard, and G. Roger. 1982. Experimental test of Bell's inequalities using time-varying analyzers. *Physical Review Letters* 49: 1804–1807.

Bell, J. 1964. On the Einstein-Podolski-Rosen paradox. *Physics* 1: 195–200. Reprinted in his 1987. *Speakable and unspeakable in quantum mechanics*. Cambridge: Cambridge University Press.

Bell, J. 1975. The theory of local beables. Technical report, CERN: TH-2053. Reprinted in his 1987. *Speakable and unspeakable in quantum mechanics*. Cambridge: Cambridge University Press.

Belnap, N., and L. Szabó. 1996. Branching space-time analysis of the GHZ theorem. *Foundations of Physics* 26(8): 982–1002.

Billingsley, P. 1995. *Probability and Measure*, 3rd edn. New York: Wiley, Inc.

Grasshoff, G., S. Portmann, and A. Wüthrich. 2005. Minimal assumption derivation of a Bell-type inequality. *British Journal for the Philosophy of Science* 56: 663–680. Preprint: quant-ph/0312176.

Hofer-Szabó, G. 2008. Separate-versus common-common-cause-type derivations of the Bell inequalities. *Synthese* 163(2): 199–215.

Hofer-Szabó, G. 2011. Bell(δ) inequalities derived from separate common causal explanation of almost perfect EPR anticorrelations. *Foundations of Physics* 41: 1398–1413.

Hofer-Szabó, G. 2012. Separate common causal explanation and the Bell inequalities. *International Journal of Theoretical Physics* 51(1): 110–123.

Hofer-Szabó, G., M. Rédei, and L. Szabó. 1999. On Reichenbach's common cause principle and Reichenbach's notion of common cause. *British Journal for the Philosophy of Science* 50: 377–399.

Norsen, T. 2006. J.S. Bell's concept of local causality. Arxiv preprint quant-ph/0601205.

Reichenbach, H. 1956. *The direction of time*. Berkeley: University of California Press.

Seevinck, M., and J. Uffink. 2011. Not throwing out the baby with the bathwater: Bell's condition of local causality mathematically 'sharp and clean'. In *Explanation, prediction, and confirmation: New trends and old ones reconsidered*, ed. D. Dieks, W. Gonzalez, S. Hartmann, T. Uebel, and M. Weber, 425–450. Dordrecht: Springer.

Szabó, L. E. 2000. Attempt to resolve the EPR-Bell paradox via Reichenbach's concept of common cause. *International Journal of Theoretical Physics* 39: 901–911.

van Fraassen, B. 1982. The Charybdis of realism: Implications of Bell's inequality. *Synthese* 52: 25–38. Reprinted in 1989. *Philosophical consequences of quantum theory: Reflections on Bell's theorem*, ed. J. T. Cushing, and E. McMullin. Notre Dame: University of Notre Dame Press.

Chapter 6
Small Probability Space Formulation of Bell's Theorem

Márton Gömöri and Tomasz Placek

To László E. Szabó

Abstract A small probability space representation of quantum mechanical probabilities is defined as a collection of Kolmogorovian probability spaces, each of which is associated with a context of a maximal set of compatible measurements, that portrays quantum probabilities as Kolmogorovian probabilities of classical events. Bell's theorem is stated and analyzed in terms of the small probability space formalism.

6.1 Introduction

There are two remarkably different ways in which Bell's inequalities can be obtained. Following Bell's original reasoning, the standard approach is to derive Bell's inequalities from the requirement of a common causal explanation for a certain set of events and their probabilities (Bell 1964, 1976; Clauser and Shimony 1978; Hofer-Szabó et al. 2013, Ch. 9).[1] The other approach, based on the results of Pitowsky and Fine, does not invoke causal notions but derives Bell's inequalities

---

[1]Bell's original 1964 derivation proceeds in two steps: first, he refers to the EPR argument as an implication from certain predictions of quantum mechanics to deterministic hidden variables, and second, he shows that deterministic hidden variables imply the Bell inequalities. Despite appearances, as is argued by Gömöri and Hofer-Szabó (2017), the EPR argument is essentially the expression of the need for a common casual explanation for the EPR correlations with EPR's famous Criterion of Reality being a special case of the Common Cause Principle.

M. Gömöri (✉)
Institute of Philosophy, Research Centre for the Humanities, Hungarian Academy of Sciences, Budapest, Hungary
e-mail: gomorim@gmail.com

T. Placek
Jagiellonian University, Kraków, Poland

© Springer International Publishing AG 2017   109
G. Hofer-Szabó, L. Wroński (eds.), *Making it Formally Explicit*, European Studies in Philosophy of Science 6, DOI 10.1007/978-3-319-55486-0_6

from the requirement of a certain set of probability-like numbers being representable in a Kolmogorovian probability space (Fine 1982a,b; Accardi 1984; Pitowsky 1989).

Bell's inequalities are famously violated by a certain set of quantum mechanical probabilities that have been measured and confirmed in various experiments. The standard approach and the Pitowsky–Fine approach differ remarkably in what this fact is taken to express. According to the standard approach the violation of Bell's inequalities implies that there is no common causal explanation (of the required sort) for events described by the quantum mechanical probabilities in question. According to the Pitowsky–Fine approach the violation of Bell's inequalities implies that there are no events representable in a Kolmogorovian probability space whose probabilities would be equal to the quantum mechanical probabilities in question.

There is an apparent tension between these two conclusions. If, according to the Pitowsky–Fine approach, there are no events representable in a Kolmogorovian probability space whose probabilities are equal to the quantum mechanical probabilities in question, then *what sort of events and what sort of probabilities attached to them* are required to be explainable in terms of common causes along the lines of the standard approach? Before answering this question the standard version of Bell's theorem cannot be meaningfully formulated.

This point was first raised by Szabó (1995). His reading of the violation of Bell's inequalities à la Pitowsky–Fine is that there cannot exist events in the world whose probabilities are equal to the quantum mechanical probabilities in question. However, there is a sense in which quantum probabilities reflect the probabilities of real events: Szabó's proposal is that quantum probabilities are *conditional* in nature. That is, whenever we say $\langle \psi, P_A \psi \rangle$ we do not refer to a probability $p(A)$ of an event A, but rather to a conditional probability $p(A \mid a)$, which is to be read as the probability of a measurement outcome A *given* that the corresponding measurement a has been performed. This understanding of quantum mechanical probabilities is in harmony with the laboratory practice of testing the predictions of quantum mechanics.

In accordance with this interpretation, Szabó proves that there exists a large Kolmogorovian probability space Σ in which the quantum mechanical probabilities that violate Bell's inequalities can be represented *as conditional probabilities*. This space is large insofar as it not only contains events representing outcomes of measurements but also events representing measurement settings. It is in terms of this probability space Σ that the requirement of common casual explanation can be meaningfully formulated and the standard version of Bell's theorem can be stated. For, according to Szabó's proposal, the events and probabilities to be explained reside in probability space Σ.

In the present paper we suggest a slightly different framework for the formulation of Bell's theorem. The philosophy is taken from Szabó: quantum probabilities are not absolute probabilities but probabilities relativized to different measurements. However, instead of starting out from one large probability space Σ we shall use a collection of small probability spaces $\{\Sigma_i\}$ each of which is associated with a certain measurement context.[2] The motivation for this is the following. The large probability space Σ includes the measurement settings and their probabilities. These data may vary from experiment to experiment, so we have a different Σ space for each instance of the kind of experiment in question. What all these different Σ spaces will agree on—and this has been confirmed in experiments of diverse sort—are the conditional probabilities representing the quantum mechanical probabilities. It is thus reasonable to see those conditional probabilities, the robust part of the data, as encoding inherent physical properties of the system in question, as opposed to the probabilities of measurement settings, the varying part of the data, determined by conditions coming from outside of the system, for example by the experimenter's decisions. In this sense the probabilities of measurement settings correspond to an "excess structure" of Σ that is not inherent to quantum mechanical phenomena and that we do not expect to be accounted for by our common causal explanation. It is when this sort of excess structure is factored out from Szabó's large probability space that we arrive at the small probability space models of quantum mechanical probabilities.[3]

The paper is structured as follows. In Sect. 6.2 we formulate the precise definitions of representations of quantum probabilities in large versus small probability spaces and investigate their relationship. In Sect. 6.3 we construct the representation of quantum probabilities associated with the Bell–Aspect experiment in small probability spaces. To our knowledge, this has never been explicitly done in the literature (although the topic of large versus small probability spaces for quantum experiments has been discussed, see Fine 1982a,b; Butterfield 1992; Svetlichny et al. 1988; Müller and Placek 2001). In Sect. 6.4 we characterize the requirement of common causal explanation as the existence of a hidden variable model consisting of a further collection of small probability spaces. We will describe these spaces for the Bell–Aspect experiment. Finally, in Sect. 6.5 we formulate the standard version of Bell's theorem and connect it with the Pitowsky–Fine version in terms of small probability spaces. More specifically, we demonstrate that the three standard premises of Bell's theorem, Parameter Independence (PI), Outcome Independence (OI), and No Conspiracy (NOCONS), permit one to paste together the small probability spaces representing the Bell–Aspect experiment into one Kolmogorovian probability space, a Kolmogorovian probability representation of this experiment in the original sense of Pitowsky–Fine. Since the latter is not to be had (because of the violation of Bell's inequalities and Pitowsky's theorem), OI, PI, and NOCONS cannot be jointly true.

---

[2]The distinction of small vs. large probability spaces was introduced in the context of Bell's theorem in Butterfield (1992), which also contains the first discussion of the corresponding two kinds of models for quantum correlations.

[3]Note that this kind of factorization by no means involves assumption about the autonomy of measurement choice in the sense of no-conspiracy. This assumption will be expressed in terms of the small probability space models in Sect. 6.4.

6.2   Representing Quantum Probabilities

Definition 1 A *quantum mechanical experiment* is a triple $\langle \mathcal{H}, \psi, \mathcal{O}_n \rangle$ where $\mathcal{H}$ is a Hilbert space, $\psi \in \mathcal{H}$ is a unit vector, and $\mathcal{O}_n = \{P_1, \ldots, P_n\}$ is a set of n projection operators on $\mathcal{H}$.[4]

A quantum mechanical experiment represents a finite collection of "yes/no" type quantum mechanical measurements performed in a given quantum state. By "quantum probabilities" we mean the numbers delivered by the quantum mechanical statistical algorithm:

$$p_i = \langle \psi, P_i \psi \rangle$$

and

$$p_{ij\ldots k} = \langle \psi, P_i P_j \ldots P_k \psi \rangle$$

whenever $P_i, P_j, \ldots, P_k$ pairwise commute. Such a data set of quantum probabilities $\langle p_1, \ldots, p_n, \ldots, p_{ij\ldots k}, \ldots \rangle$ will be referred to as the *correlation vector* associated with the quantum mechanical experiment $\langle \mathcal{H}, \psi, \mathcal{O}_n \rangle$ in question.

A correlation vector associated with a quantum mechanical experiment consists of probability-like numbers. But are they really probabilities? Are there events in the world whose probabilities are equal to quantum probabilities? More precisely, are there events occurring in the runs of a real-world experiment whose relative frequencies are equal to quantum probabilities? A minimal condition for this to be the case is that one has to be able to construct a Kolmogorovian probability space in which those events and their probabilities are accommodated.[5]

Definition 2 A *Kolmogorovian representation* of a quantum mechanical experiment $\langle \mathcal{H}, \psi, \mathcal{O}_n \rangle$ is a Kolmogorovian probability space $\langle \Omega, \mathcal{F}, p \rangle$ that satisfies the following conditions:

- to each $P_i \in \mathcal{O}_n$ there corresponds an $\tilde{A}_i \in \mathcal{F}$ such that

$$p\left(\tilde{A}_i\right) = \langle \psi, P_i \psi \rangle \qquad (6.1)$$

- for all $P_i, P_j \ldots, P_k \in \mathcal{O}_n$, if they pairwise commute, then

$$p\left(\tilde{A}_i \cap \tilde{A}_j \cap \ldots \cap \tilde{A}_k\right) = \langle \psi, P_i P_j \ldots P_k \psi \rangle \qquad (6.2)$$

---

[4]This definition is taken from Feintzeig (2015).
[5]For more on this point see Szabó's (2001) Laboratory Record Argument.

As we know from Pitowsky (1989), a quantum mechanical experiment whose associated correlation vector violates the appropriate Bell-type inequalities does not have a Kolmogorovian representation. The correlation vector

$$\langle p_1, p_2, p_3, p_4, p_{13}, p_{14}, p_{23}, p_{24} \rangle = \left\langle \frac{1}{2}, \frac{1}{2}, \frac{1}{2}, \frac{1}{2}, \frac{3}{8}, \frac{3}{8}, 0, \frac{3}{8} \right\rangle \tag{6.3}$$

associated with the Bell–Aspect experiment does violate the appropriate Bell-type inequalities. Consequently:

Theorem 1 *There is a quantum mechanical experiment with no Kolmogorovian representation.*

Thus, quantum probabilities are not Kolmogorovian in general. If not Kolmogorovian, then what sort of probabilities are they?

According to Szabó's (1995, 2001) proposal quantum probabilities are Kolmogorovian *conditional* probabilities.

Definition 3 A *large conditional Kolmogorovian representation* of a quantum mechanical experiment $\langle \mathcal{H}, \psi, \mathcal{O}_n \rangle$ is a Kolmorogovian probability space $\langle \Omega, \mathcal{F}, p \rangle$ that satisfies the following conditions:

- to each $P_i \in \mathcal{O}_n$ there corresponds two events $A_i, a_i \in \mathcal{F}, A_i \subseteq a_i$ such that

$$p(A_i \mid a_i) = \langle \psi, P_i \psi \rangle \tag{6.4}$$

- for all $P_i, P_j \in \mathcal{O}_n$, if they do not commute, then $a_i \cap a_j = \emptyset$
- for all $P_i, P_j \ldots, P_k \in \mathcal{O}_n$, if they pairwise commute, then $p(a_i \cap a_j \cap \ldots \cap a_k) > 0$ and

$$p\left(A_i^\pm \cap A_j^\pm \cap \ldots \cap A_k^\pm \mid a_i \cap a_j \cap \ldots \cap a_k\right) = \langle \psi, P_i^\pm P_j^\pm \ldots P_k^\pm \psi \rangle \tag{6.5}$$

where $A_i^+ = A_i, A_i^- = a_i \setminus A_i, P_i^+ = P_i, P_i^- = P_i^\perp; P_i^\perp$ is the orthogonal complement of P_i.

Here a_i is to be read as the event of setting up a measurement device to perform a certain "yes/no" type measurement, and A_i as the event of the firing of the corresponding "yes" detector. The "incompatibility" of two measurements is coded in the postulate that the corresponding two measurement settings are disjoint.

Since the large probability space $\langle \Omega, \mathcal{F}, p \rangle$ is Kolmogorovian, the probabilities of events in it can be interpreted as relative frequencies measured in a real-world experiment. In contrast, the values of conditional probabilities (6.4) and (6.5) that stand for the quantum mechanical probabilities may well turn out not to have a Kolmogorovian representation since they are conditional probabilities *with conditioning on different events* (as opposed to conditioning on one single event). The values of conditional probabilities pertaining to different conditions typically cannot be delivered (as absolute probabilities) by a single Kolmogorovian probability measure. This explains why correlation vectors like (6.3), associated with the Bell–Aspect experiment, do not have a Kolmogorovian representation. This is because (6.3) is in fact a collection of conditional probabilities

$$\langle p\,(A_1 \mid a_1)\,, p\,(A_2 \mid a_2)\,, p\,(A_3 \mid a_3)\,, p\,(A_4 \mid a_4)\,, p\,(A_1 \cap A_3 \mid a_1 \cap a_3)\,,$$

$$p\,(A_1 \cap A_4 \mid a_1 \cap a_4)\,, p\,(A_2 \cap A_3 \mid a_2 \cap a_3)\,, p\,(A_2 \cap A_4 \mid a_2 \cap a_4)\rangle$$

that do not belong to the same condition.

Now we formulate an alternative characterization of the sense in which quantum probabilities can be regarded as "conditional".

Definition 4 A *small Kolmogorovian representation* of a quantum mechanical experiment $\langle \mathcal{H}, \psi, \mathcal{O}_n \rangle$ is a finite collection of Kolmogorovian probability spaces $\{\langle \Omega_\alpha, \mathcal{F}_\alpha, p_\alpha \rangle\}_{\alpha \in I}$ that satisfies the following conditions:

- for $\alpha \neq \beta$, $\Omega_\alpha \cap \Omega_\beta = \emptyset$
- to each maximal set of pairwise commuting projection operators $\mathcal{O} \subseteq \mathcal{O}_n$ there corresponds a Kolmogorovian probability space $\langle \Omega_\alpha, \mathcal{F}_\alpha, p_\alpha \rangle$ ($\alpha \in I$) such that $\langle \Omega_\alpha, \mathcal{F}_\alpha, p_\alpha \rangle$ is a Kolmogorovian representation of the quantum mechanical experiment $\langle \mathcal{H}, \psi, \mathcal{O} \rangle$

Here a probability space $\langle \Omega_\alpha, \mathcal{F}_\alpha, p_\alpha \rangle$ represents the probabilities of measurement outcomes in a certain measurement context. The contexts are provided by the maximal conjunctions of measurements in the quantum mechanical experiment in question which are (and *a fortiori* can be) performed simultaneously. In other words, $\langle \Omega_\alpha, \mathcal{F}_\alpha, p_\alpha \rangle$ encodes the relative frequencies of outcomes in a real-world experiment over the subensemble of experimental runs where a given maximal conjunction of measurements is performed.

Although each small probability space $\langle \Omega_\alpha, \mathcal{F}_\alpha, p_\alpha \rangle$ alone is Kolmogorovian, quantum probabilities pertaining to different maximal sets of compatible measurements reside in different members of the collection $\{\langle \Omega_\alpha, \mathcal{F}_\alpha, p_\alpha \rangle\}_{\alpha \in I}$. Probability values belonging to different Kolmogorovian probability spaces do not make up a Kolmogorovian probability measure in general. This is another way to see why correlation vectors like (6.3) do not have a Kolmogorovian representation. It is because (6.3) is in fact a collection of probabilities

$$\langle p_{1j}\,(A_1)\,, p_{2j}\,(A_2)\,, p_{i3}\,(A_3)\,, p_{i4}\,(A_4)\,,$$

$$p_{13}\,(A_1 \cap A_3)\,, p_{14}\,(A_1 \cap A_4)\,, p_{23}\,(A_2 \cap A_3)\,, p_{24}\,(A_2 \cap A_4)\rangle$$

that come from different probability spaces—labeled by the pairs of compatible measurements ($i = 1, 2; j = 3, 4$).[6]

---

[6]The idea of small Kolmogorovian representation goes back to Kolmogorov (1933) at least, who argued that each experimental arrangement generates its own probability space. Kolmogorov was clear that one should not expect that probabilities of outcomes obtained in various different

A large conditional Kolmogorovian representation portrays quantum probabilities as conditional probabilities in a large probability space. A small Kolmogorovian representation portrays quantum probabilities as absolute probabilities in many small probability spaces. We shall now demonstrate that large conditional and small Kolmogorovian representations are mutually translatable, though the translation from many small spaces to one large space requires some extra data. First we show that there is a natural translation between large and small Kolmogorovian probability spaces.

Proposition 1 *Let $\langle \Omega, \mathcal{F}, p \rangle$ be a Kolmogorovian probability space and $\{\Omega_\alpha\}_{\alpha \in I} \subseteq \mathcal{F}$ such that $p(\Omega_\alpha) > 0$. Define $\mathcal{F}_\alpha := \{X \subseteq \Omega_\alpha \mid X \in \mathcal{F}\}$ and $p_\alpha(X) := p(X \mid \Omega_\alpha)$ for $X \in \mathcal{F}_\alpha$. Then $\{\langle \Omega_\alpha, \mathcal{F}_\alpha, p_\alpha \rangle\}_{\alpha \in I}$ is a collection of Kolmogorovian probability spaces.*

Proof It is immediate to see that $\mathcal{F}_\alpha$ is a field of subsets of Ω_α. As for the satisfaction of Kolmogorov's axioms, clearly $p_\alpha(\Omega_\alpha) = 1$. For the additivity axiom, let $\mathcal{G}$ be a countable family of pairwise disjoint elements of $\mathcal{F}_\alpha$; then $p_\alpha(\cup \mathcal{G}) = p(\cup \mathcal{G} \mid \Omega_\alpha) = p(\cup \mathcal{G} \cap \Omega_\alpha)/p(\Omega_\alpha) = \sum_{A \in \mathcal{G}} p(A \cap \Omega_\alpha)/p(\Omega_\alpha) = \sum_{A \in \mathcal{G}} p(A \mid \Omega_\alpha) = \sum_{A \in \mathcal{G}} p_\alpha(A)$. □

Proposition 2 *Let $\{\langle \Omega_\alpha, \mathcal{F}_\alpha, p_\alpha \rangle\}_{\alpha \in I}$ be a finite collection of Kolmogorovian probability spaces and $\{c_\alpha\}_{\alpha \in I}$ a set of positive coefficients that sum up to 1. Then $\langle \Omega, \mathcal{F}, p \rangle$ is a Kolmogorovian probability space, where*

- $\Omega = \bigcup_{\alpha \in I} \Omega_\alpha$
- for $E \subseteq \Omega$, $E \in \mathcal{F}$ iff $E \cap \Omega_\alpha \in \mathcal{F}_\alpha$
- for $E \in \mathcal{F}$, $p(E) = \sum_{\alpha \in I} c_\alpha \cdot p_\alpha(E \cap \Omega_\alpha)$

Proof It is straightforward to see that $\mathcal{F}$ is a field, indeed. To check the axioms, let us calculate first $p(\Omega) = \sum_{\alpha \in I} c_\alpha \cdot p_\alpha(\Omega \cap \Omega_\alpha) = \sum_{\alpha \in I} c_\alpha \cdot p_\alpha(\Omega_\alpha) = \sum_{\alpha \in I} c_\alpha = 1$. Next, let $\mathcal{G}$ be a countable family of pairwise disjoint elements of $\mathcal{F}$, $p(\cup \mathcal{G}) = \sum_{\alpha \in I} c_\alpha \cdot p_\alpha(\cup \mathcal{G} \cap \Omega_\alpha) = \sum_{\alpha \in I} c_\alpha \cdot p_\alpha(\bigcup_{A \in \mathcal{G}}(A \cap \Omega_\alpha)) = \sum_{\alpha \in I} \sum_{A \in \mathcal{G}} c_\alpha p_\alpha(A \cap \Omega_\alpha) = \sum_{A \in \mathcal{G}} \sum_{\alpha \in I} c_\alpha p_\alpha(A \cap \Omega_\alpha) = \sum_{A \in \mathcal{G}} p(A)$.

Note that the original ("p_α") probabilities are returned by conditionalization: if X belongs to $\mathcal{F}_\alpha$, then $p_\alpha(X) = p(X|\Omega_\alpha) = p(X)/p(\Omega_\alpha) = c_\alpha p_\alpha(X)/c_\alpha = p_\alpha(X)$.[7] □

This natural translation allows us to construct small Kolmogorovian representations from large ones, and vica versa.

Proposition 3 *A quantum mechanical experiment has a large conditional Kolmogorovian representations iff it has a small Kolmogorovian representation.*

---

experimental arrangements can be represented, as absolute probabilities, in a single (perhaps huge) probability space (Khrennikov 2010, p. 26).

[7] Thanks to Zalán Gyenis for discussions on this topic.

Proof (a) Suppose that $\langle \Omega, \mathcal{F}, p \rangle$ is a large conditional Kolmogorovian representation of $\langle \mathcal{H}, \psi, \mathcal{O}_n \rangle$. Consider the collection of Kolmogorovian probability spaces $\{\langle \Omega_\alpha, \mathcal{F}_\alpha, p_\alpha \rangle\}_{\alpha \in I}$ constructed from $\langle \Omega, \mathcal{F}, p \rangle$ in Proposition 1 with $\Omega_\alpha = a_m \cap a_n \cap \ldots \cap a_s$. Here, following the notation of Definition 3, $a_m \cap a_n \cap \ldots \cap a_s$ corresponds to a maximal conjunction of compatible measurements in $\langle \mathcal{H}, \psi, \mathcal{O}_n \rangle$ as represented in large conditional Kolmogorovian representation $\langle \Omega, \mathcal{F}, p \rangle$; $\alpha \in I$ runs over the set of all such conjunctions.

Consider a given $\alpha \in I$. Let $\mathcal{O}_\alpha = \{P_m, P_n \ldots, P_s\} \subseteq \mathcal{O}_n$ be the corresponding maximal set of commuting projectors. We demonstrate that $\langle \Omega_\alpha, \mathcal{F}_\alpha, p_\alpha \rangle$ is a Kolmogorovian representation of $\langle \mathcal{H}, \psi, \mathcal{O}_\alpha \rangle$. Let $\tilde{A}_i = A_i \cap \Omega_\alpha = A_i \cap a_m \cap a_n \cap \ldots \cap a_s$ where $A_i \in \mathcal{F}$ is the "outcome" set representing $P_i \in \mathcal{O}_\alpha$ in $\langle \Omega, \mathcal{F}, p \rangle$. Obviously, $\tilde{A}_i \in \mathcal{F}_\alpha$. The following shows that $\tilde{A}_i$ is a representation of P_i in $\langle \Omega_\alpha, \mathcal{F}_\alpha, p_\alpha \rangle$ in the sense of (6.1):

$$p_\alpha \left(\tilde{A}_i \right) = p \left(\tilde{A}_i \mid \Omega_\alpha \right) = p \left(A_i \cap \Omega_\alpha \mid \Omega_\alpha \right)$$

$$= p \left(A_i \mid a_i \cap a_j \cap \ldots \cap a_k \right) \overset{\star}{=} p \left(A_i \mid a_i \right) = \langle \psi, P_i \psi \rangle$$

Further, let $P_i, P_j, \ldots, P_k \in \mathcal{O}_\alpha$. Then:

$$p_\alpha \left(\tilde{A}_i \cap \tilde{A}_j \cap \ldots \cap \tilde{A}_k \right) = p \left(\tilde{A}_i \cap \tilde{A}_j \cap \ldots \cap \tilde{A}_k \mid \Omega_\alpha \right)$$

$$= p \left(A_i \cap A_j \cap \ldots \cap A_k \cap \Omega_\alpha \mid \Omega_\alpha \right)$$

$$= p \left(A_i \cap A_j \cap \ldots \cap A_k \mid a_m \cap a_n \cap \ldots \cap a_s \right)$$

$$\overset{\star}{=} p \left(A_i \cap A_j \cap \ldots \cap A_k \mid a_i \cap a_j \cap \ldots \cap a_k \right)$$

$$= \langle \psi, P_i P_j \ldots P_k \psi \rangle$$

that is (6.2) is also satisfied.

In the equalities with $\star$ we made use of the fact that "the statistics of outcomes of a given set of measurements is not altered by performing a further measurement". This is a consequence of the fact that in a commutative sublattice of a Hilbert lattice the law of total probability holds. For example, for all $P_i, P_j \in \mathcal{O}_\alpha$:

$$p \left(A_i \mid a_i \right) = \langle \psi, P_i \psi \rangle = \langle \psi, P_i P_j \psi \rangle + \langle \psi, P_i P_j^\perp \psi \rangle$$

$$= p \left(A_i \cap A_j \mid a_i \cap a_j \right) + p \left(A_i \cap \left(a_j \setminus A_j \right) \mid a_i \cap a_j \right)$$

$$= p \left(A_i \mid a_i \cap a_j \right)$$

The derivation is similar for more complex conjunctions of measurements.

(b) Suppose that $\{\langle \Omega_\alpha, \mathcal{F}_\alpha, p_\alpha \rangle\}_{\alpha \in I}$ is a small Kolmogorovian representation of $\langle \mathcal{H}, \psi, \mathcal{O}_n \rangle$. Consider the Kolmogorovian probability space $\langle \Omega, \mathcal{F}, p \rangle$ constructed from $\{\langle \Omega_\alpha, \mathcal{F}_\alpha, p_\alpha \rangle\}_{\alpha \in I}$ in Proposition 2 with an arbitrary set of positive coefficients

$\{c_\alpha\}_{\alpha \in I}$ that sum up to 1. We demonstrate that $\langle \Omega, \mathcal{F}, p \rangle$ is a large conditional Kolmogorovian representation of $\langle \mathcal{H}, \psi, \mathcal{O}_n \rangle$.

First, consider a $P_i \in \mathcal{O}_n$. Let

$$A_i = \bigcup_{\alpha : P_i \in \mathcal{O}_\alpha} A_i^\alpha$$

$$a_i = \bigcup_{\alpha : P_i \in \mathcal{O}_\alpha} \Omega_\alpha$$

where $A_i^\alpha \in \mathcal{F}_\alpha$ is the representing set of P_i in $\langle \Omega_\alpha, \mathcal{F}_\alpha, p_\alpha \rangle$. Obviously, $A_i, a_i \in \mathcal{F}$. We show that A_i, a_i provide a "conditional representation" of P_i in $\langle \Omega, \mathcal{F}, p \rangle$ in the sense that (6.4) is satisfied.

$$p(A_i \mid a_i) = p\left(\bigcup_{\alpha : P_i \in \mathcal{O}_\alpha} A_i^\alpha \ \Bigg| \ \bigcup_{\alpha : P_i \in \mathcal{O}_\alpha} \Omega_\alpha \right) = \frac{p\left(\bigcup_{\alpha : P_i \in \mathcal{O}_\alpha} A_i^\alpha \right)}{p\left(\bigcup_{\alpha : P_i \in \mathcal{O}_\alpha} \Omega_\alpha \right)}$$

As for the numerator,

$$p\left(\bigcup_{\alpha : P_i \in \mathcal{O}_\alpha} A_i^\alpha \right) = \sum_{\beta \in I} c_\beta \cdot p_\beta \left(\bigcup_{\alpha : P_i \in \mathcal{O}_\alpha} A_i^\alpha \cap \Omega_\beta \right)$$

$$\overset{\star\star}{=} \sum_{\alpha : P_i \in \mathcal{O}_\alpha} c_\alpha \cdot p_\alpha \left(A_i^\alpha \right) = \left(\sum_{\alpha : P_i \in \mathcal{O}_\alpha} c_\alpha \right) \cdot \langle \psi, P_i \psi \rangle$$

as for the denominator,

$$p\left(\bigcup_{\alpha : P_i \in \mathcal{O}_\alpha} \Omega_\alpha \right) = \sum_{\beta \in I} c_\beta \cdot p_\beta \left(\bigcup_{\alpha : P_i \in \mathcal{O}_\alpha} \Omega_\alpha \cap \Omega_\beta \right)$$

$$\overset{\star\star}{=} \sum_{\alpha : P_i \in \mathcal{O}_\alpha} c_\alpha \cdot p_\alpha (\Omega_\alpha) = \sum_{\alpha : P_i \in \mathcal{O}_\alpha} c_\alpha$$

Hence, indeed:

$$p(A_i \mid a_i) = \langle \psi, P_i \psi \rangle$$

In the equalities with $\star\star$ we used that in a small Kolmogorovian representation the Ω_α-s are disjoint.

Second, if $P_i, P_j \in \mathcal{O}_n$ do not commute then there is no $\alpha \in I$ such that $P_i, P_j \in \mathcal{O}_\alpha$. Since the Ω_α-s are disjoint, then indeed $a_i \cap a_j = \emptyset$ as required in Definition 3.

Third, suppose that $P_i, P_j \ldots, P_k \in \mathcal{O}_n$ pairwise commute. Then there is an $\alpha \in I$ such that $\Omega_\alpha \subseteq a_i \cap a_j \cap \ldots \cap a_k$ and hence

$$p\left(a_i \cap a_j \cap \ldots \cap a_k\right) \geqslant p\left(\Omega_\alpha\right) = \sum_{\beta \in I} c_\beta \cdot p_\beta \left(\Omega_\alpha \cap \Omega_\beta\right) = c_\alpha \cdot p_\alpha \left(\Omega_\alpha\right) = c_\alpha > 0$$

Further, a similar calculation as the single outcome case above yields

$$p\left(A_i^\pm \cap A_j^\pm \cap \ldots \cap A_k^\pm \mid a_i \cap a_j \cap \ldots \cap a_k\right) = \left\langle \psi, P_i^\pm P_j^\pm \ldots P_k^\pm \psi \right\rangle$$

That is, (6.5) is meaningful and satisfied. Note that in this calculation one has to make use of the fact that in a Kolmogorovian representation $\langle \Omega, \mathcal{F}, p \rangle$ not only (6.2) but also

$$p\left(\tilde{A}_i^\pm \cap \tilde{A}_j^\pm \cap \ldots \cap \tilde{A}_k^\pm\right) = \left\langle \psi, P_i^\pm P_j^\pm \ldots P_k^\pm \psi \right\rangle$$

holds, where $\tilde{A}_i^+ = \tilde{A}_i, \tilde{A}_i^- = \Omega \setminus \tilde{A}_i$.   □

It is a well-known fact of Hilbert space quantum mechanics that a commutative sublattice of a Hilbert lattice is Boolean; further, quantum probability restricted to a Boolean sublattice forms a Kolmogorovian probability measure. This implies that a quantum mechanical experiment $\langle \mathcal{H}, \psi, \mathcal{O} \rangle$ in which all projection operators of $\mathcal{O}$ pairwise commute has a Kolmogorovian representation. This of course remains true when $\mathcal{O}$ is a maximal subset of pairwise commuting projection operators of a set of projectors $\mathcal{O}_n$. Therefore, an arbitrary quantum mechanical experiment $\langle \mathcal{H}, \psi, \mathcal{O}_n \rangle$ has a small Kolmogorovian representation. From Proposition 3 it then follows that it also has a large conditional Kolmogorovian representation. Thus:

Theorem 2 *All quantum mechanical experiments have large conditional and small Kolmogorovian representations.*

That is to say, quantum probabilities, as conditional probabilities in a large space or as absolute probabilities in small spaces, are always Kolmogorovian. This claim can be taken as a version of what Szabó (1995) calls the *Kolmogorovian Censorship*.

The only difference between large conditional and small Kolmogorovian representations is the extra data of coefficients $\{c_\alpha\}_{\alpha \in I}$ required for building a large space from the collection of small ones. This data determine the probabilities of the measurement settings. In what follows we shall focus on small representations that do not have this excess structure.

The rest of the paper is restricted to the discussion of the Bell–Aspect experiment.

6.3   Small Kolmogorovian Representation of the Bell–Aspect Experiment

Now we construct a small Kolmogorovian representation of the Bell–Aspect experiment. Consider four non-empty sets a_1, a_2, b_3, b_4 representing measurement settings on the left (a-settings) and on the right (b-settings), and subsets $A_i^+, A_i^- \subseteq a_i, B_j^+, B_j^- \subseteq b_j$ to be thought of as measurement outcomes of a_i and b_j, respectively. We shall use the convention $i \in \{1, 2\}$ and $j \in \{3, 4\}$, and $m, n \in \{-, +\}$. We assume that these sets satisfy the following:

$$a_1 \cap a_2 = b_3 \cap b_4 = \emptyset$$
$$a_i \cap b_j \neq \emptyset$$
$$a_1 \cup a_2 = b_3 \cup b_4$$
$$A_i^- \cup A_i^+ = a_i, \quad B_j^- \cup B_j^+ = b_j \tag{6.6}$$
$$A_i^- \cap A_i^+ = \emptyset, \quad B_j^- \cap B_j^+ = \emptyset$$
$$\text{for } i = 1, 2; j = 3, 4$$

These set-theoretical requirements express conditions deriving from the physical meaning of the events in question; e.g. that in each run of the experiment a setting on the left and a setting on the right is to be selected, or that A_i^m cannot occur together with A_i^n if $m \neq n$. Out of these sets we build now four Kolmogorovian probability spaces, labeled by pairs $\langle i, j \rangle$, in which quantum mechanical probabilities will be represented. We exhibit below the probability space $\langle \Omega_{ij}, \mathcal{F}_{ij}, p_{ij} \rangle$:

$$\Omega_{ij} = a_i \cap b_j$$
$$\mathcal{F}_{ij} = \{\emptyset, A_i^+ \cap B_j^+, A_i^- \cap B_j^+, A_i^+ \cap B_j^-, A_i^- \cap B_j^-,$$
$$A_i^+ \cap b_j, A_i^- \cap b_j, a_i \cap B_j^+, a_i \cap B_j^-, \Omega_{ij}\} \tag{6.7}$$
$$p_{ij}(A_i^m \cap b_j) = \langle \psi, P_i^m \psi \rangle \quad p_{ij}(a_i \cap B_j^n) = \langle \psi, P_j^n \psi \rangle$$
$$p_{ij}(A_i^m \cap B_j^n) = \langle \psi, P_i^m P_j^n \psi \rangle$$

where P_i^m and P_j^n are projections corresponding to outcomes A_i^m and B_j^n, respectively, and ψ is the singlet state used in the Bell–Aspect experiment. A particular arrangement of the measurement settings yields a probability measure p_{ij} that corresponds to correlation vector (6.3).

Surface locality, according to which the probability of a measurement outcome in one wing does not depend on the choice of a measurement setting in the other wing, follows from the fact the measures p_{ij} represent quantum probabilities in line with (6.7). Observe the shape of this condition in the small space formalism:

$$p_{ij}(A_i^m \cap b_j) = p_{ij'}(A_i^m \cap b_{j'}) \text{ and } p_{ij}(a_i \cap B_j^n) = p_{i'j}(a_{i'} \cap B_j^n)$$

The condition relates different probability spaces, as it refers to probability measures belonging to different spaces. In what follows we will come across more constraints relating different small probability spaces.

6.4 Hidden Variables: Creation of New Small Probability Spaces

The four Kolmogorovian probability spaces we constructed above offer a probabilistic representation of surface data: of measurement outcomes and their probabilities. In the Bell–Aspect experiment these probabilities are such that there are statistical correlations between the outcomes in the two wings. It is easy to read off from the numbers provided by correlation vector (6.3) that

$$p_{ij}\left(\left(A_i^m \cap b_j\right) \cap \left(a_i \cap B_j^n\right)\right) \neq p_{ij}\left(A_i^m \cap b_j\right) p_{ij}\left(a_i \cap B_j^n\right)$$

Assuming the spatial separation of the two wings, this correlation calls for a common causal explanation. In the hidden variable terminology, a common causal explanation translates to the requirement that the surface data, including the correlations to be explained, emerge from a deeper level of hidden states. We now produce small probability spaces that accommodate these hidden states.

The new probability spaces should deliver probabilities of measurements as marginal probability. In accord with the ideology of the common common cause (Belnap and Szabó 1996), we thus consider a set Λ, interpreted as the set of all hidden variables, one for *all* pairs of settings a_i, b_j. With Λ we produce four product probability spaces $\langle \Omega_{ij\Lambda}, \mathfrak{F}_{ij\Lambda}, \mu_{ij\Lambda} \rangle$, where[8]

$$\Omega_{ij\Lambda} = \{A_i^m \cap B_j^n \mid m, n \in \{-, +\}\} \times \Lambda$$

$\mathfrak{F}_{ij\Lambda}$ is generated by atoms $\{\langle A_i^m \cap B_j^n, \lambda \rangle\}$, where $\lambda \in \Lambda$

$$\mu_{ij\Lambda} \text{ satisfies } \sum_{\lambda \in \Lambda} \mu_{ij\lambda}(A_i^m \cap B_j^n, \lambda) = p_{ij}(A_i^m \cap B_j^n)$$

(6.8)

Observe that the new fields leave no room for some events from the fields of the surface small probability representation, like $a_i \cap b_j =: \Omega_{ij}$ or $A_i^m \cap b_j$. However, there are in $\mathfrak{F}_{ij\Lambda}$ *ersatz* objects for such events, as the following "underline" notation explains:

---

[8]To avoid eye strain, below we do not write the angle brackets and curly brackets in parentheses for a probability function.

$$\{\langle A_i^m \cap B_j^n, \lambda\rangle \mid \lambda \in \Lambda\} =: \underline{A_i^m \cap B_j^n, \Lambda} \tag{6.9}$$

$$\{\langle A_i^m \cap B_j^n, \lambda\rangle \mid n \in \{-, +\}, \lambda \in \Lambda\} =: \underline{A_i^m b_j, \Lambda} \tag{6.10}$$

$$\{\langle A_i^m \cap B_j^n, \lambda\rangle \mid m \in \{-, +\}, \lambda \in \Lambda\} =: \underline{a_i B_j^n, \Lambda} \tag{6.11}$$

$$\{\langle A_i^m \cap B_j^n, \lambda\rangle \mid m \in \{-, +\}, n \in \{-, +\}, \lambda \in \Lambda\} =: \underline{a_i b_j, \Lambda} \tag{6.12}$$

$$\{\langle A_i^m \cap B_j^n, \lambda\rangle \mid m \in \{-, +\}, n \in \{-, +\}\} =: \underline{a_i b_j, \lambda} \tag{6.13}$$

Since we need μ-probabilities conditional on λ to be well-defined, we require that Λ be countable. Otherwise, we would have $\mu_{ij\Lambda}(a_i b_j, \lambda) = 0$ for every $\lambda \in \Lambda$,[9] making conditional probabilities $\mu_{ij\Lambda}(A_i^m \cap B_j^n, \Lambda \mid \underline{a_i b_j, \lambda})$ undefined.

The introduction of Λ leads to four new product probability spaces, that could be seen as fine-graining of the initial probability spaces $\langle \Omega_{ij}, \mathcal{F}_{ij}, p_{ij}\rangle$. A large class of such product probability spaces fine-graining a given probability space can be constructed, provided these product spaces are not constrained by further conditions. Causal explanation of the results of the Bell–Aspect experiment imposes, however, such conditions, to which we now turn. We formulate them in the small-space approach.

Outcome Independence (OI) requires that the hidden variable should screen-off the correlation of outcomes registered in separate wings of the experiment. Thus, in the present formulation, it says:

$$\mu_{ij\Lambda}(\underline{A_i^m \cap B_j^n, \Lambda} \mid \underline{a_i b_j, \lambda}) = \mu_{ij\Lambda}(\underline{A_i^m b_j, \Lambda} \mid \underline{a_i b_j, \lambda}) \times \mu_{ij\Lambda}(\underline{a_i B_j^n, \Lambda} \mid \underline{a_i b_j, \lambda}) \tag{OI}$$

Parameter Independence (PI) demands that an outcome in one wing of the experiment should be statistically independent from a setting selected in the other wing given that a hidden state is specified:

$$\mu_{ij\Lambda}(\underline{A_i^m b_j, \Lambda} \mid \underline{a_i b_j, \lambda}) = \mu_{ij'\Lambda}(\underline{A_i^m b_{j'}, \Lambda} \mid \underline{a_i b_{j'}, \lambda})$$

$$\mu_{ij\Lambda}(\underline{a_i B_j^n, \Lambda} \mid \underline{a_i b_j, \lambda}) = \mu_{i'j\Lambda}(\underline{a_{i'} B_j^n, \Lambda} \mid \underline{a_{i'} b_j, \lambda}) \tag{PI}$$

Finally, No Conspiracy (NOCONS) postulates that pairs of settings be statistically independent from hidden states; in the present formalism, it reads:

$$\mu_{ij\Lambda}(\underline{a_i b_j, \lambda}) = \mu_{i'j'\Lambda}(\underline{a_{i'} b_{j'}, \lambda}) \tag{NOCONS}$$

---

[9]See theorem 10.2 in Billingsley 1995.

(Each condition OI, PI, and NOCONS is universally quantified with respect to $i, i' \in \{1, 2\}, j, j' \in \{3, 4\}, m, n \in \{-, +\}$, and $\lambda \in \Lambda$.) Note that in contrast to OI, both PI and NOCONS relate probability measures belonging to different product probability spaces.

To summarize what has been said we state the following definition.

Definition 5 A hidden-state small Kolmogorovian representation of the Bell–Aspect experiment is a quadruple of product probability spaces $\langle \Omega_{ij\Lambda}, \mathfrak{F}_{ij\Lambda}, p_{ij\Lambda} \rangle$ $(i \in \{1, 2\}, j \in \{3, 4\})$ of (6.8), with Λ countable and A_i^m, B_j^n, a_i, b_j satisfying (6.6), and p_{ij} given by (6.7).

A small-space local common causal non-conspiratorial model of the Bell–Aspect experiment is a hidden-state small Kolmogorovian representation of the Bell–Aspect experiment that satisfies the three conditions OI, PI, and NOCONS.

It is no news that the combination of OI, PI, and NOCONS permits a derivation of Bell's inequalities. It will be instructive to see, however, how this derivation goes through in a prudent framework of small product probability spaces. This will lead us to ask what mathematical meaning the combination of the three conditions has, and the answer we give will take us back to Kolmogorovian representability in the original sense of Pitowsky–Fine described in Definition 2.

6.5   The Mathematical Meaning of OI, PI, and NOCONS

To examine the mathematical meaning of OI, PI, and NOCONS, it might help to sketch a particular attitude to Bell's theorem. The essence of this attitude is a belief that working with small probability spaces, that is, by first producing a surface small Kolmogorovian representation of the Bell–Aspect experiment, and then fine-graining it to obtain a hidden-state small Kolmogorovian representation, one somehow blocks the derivation of Bell's inequalities. To some extent, this belief is justified: there is a large class of hidden-state small Kolmogorovian representations of the experiment which do not imply Bell's inequalities. But then, dramatically, once OI, PI, and NOCONS are brought into play, Bell's inequalities become derivable. So, what, mathematically speaking, do the three conditions do to a hidden-state small Kolmogorovian representation? The theorem below offers an answer to this query.

Theorem 1 *Let* $\langle \Omega_{ij\Lambda}, \mathfrak{F}_{ij\Lambda}, p_{ij\Lambda} \rangle$ *(*$i \in \{1,2\}$, $j \in \{3,4\}$*) be a hidden-state Kolmogorovian representation of the Bell–Aspect experiment, delivering surface probabilities p_{ij}. Assume that the spaces $\langle \Omega_{ij\Lambda}, \mathfrak{F}_{ij\Lambda}, p_{ij\Lambda} \rangle$ satisfy OI, PI, and NOCONS (i.e., they form a small-space model of Definition 5). Then there is a Kolmogorovian probability space $\langle \Omega, \mathfrak{F}, P \rangle$ such that all the surface probabilities p_{ij} are identifiable with absolute probabilities in $\langle \Omega, \mathfrak{F}, P \rangle$.*[10]

Proof Let $\langle \Omega_{ij\Lambda}, \mathfrak{F}_{ij\Lambda}, p_{ij\Lambda} \rangle$ ($i \in \{1,2\}$, $j \in \{3,4\}$) be as stated in the premise of the theorem. Using an idea of Fine's theorem, by applying the three conditions OI, PI, and NOCONS, we will produce a Kolmogorovian product probability space $\langle \Omega, \mathcal{F}, P \rangle$ that represents probabilities p_{ij} of the small probability spaces as absolute probabilities. The essential part of the argument is that the new probability space returns probabilities of joint and single outcomes, which is a crux of Fine's theorem.

The base set of our product space is:

$$\Omega = \{A_1^-, A_1^+\} \times \{A_2^-, A_2^+\} \times \{B_3^-, B_3^+\} \times \{B_4^-, B_4^+\}$$

its field of subsets $\mathcal{F}$ is generated by the following atoms:

$$\left\{ \langle A_1^m, A_2^n, B_3^r, B_4^t \rangle \right\}, \text{ with } m, n, r, t \in \{+, -\}$$

and the measure P is given by:

$$P(\langle A_1^m, A_2^n, B_3^r, B_4^t \rangle) =$$

$$\sum_{\lambda \in \Lambda} \mu_{13\Lambda}(\underline{a_1 b_3, \lambda}) \times \mu_{13\Lambda}(A_1^m b_3, \Lambda \mid \underline{a_1 b_3, \lambda}) \times \mu_{13\Lambda}(a_1 B_3^r, \Lambda \mid \underline{a_1 b_3, \lambda}) \times$$

$$\times \mu_{24\Lambda}(A_2^n b_4, \Lambda \mid \underline{a_2 b_4, \lambda}) \times \mu_{24\Lambda}(a_2 B_4^t, \Lambda \mid \underline{a_2 b_4, \lambda})$$

(6.14)

We need to show the satisfaction of equations like

$$P(\{\langle A_1^m, A_2^n, B_3^r, B_4^t \rangle \mid m, t \in \{-, +\}\}) = p_{23}(A_2^n \cap B_3^r)$$

We calculate only this case, observing how OI, PI, and NOCONS permit one to change labels of small probability spaces. In the calculation we indicate where these conditions are used (we leave other cases as an exercise for the reader).

---

[10] An analogous construction, but in a specific and little known framework of stochastic outcomes in branching spacetimes (SOBST) is carried out in Müller and Placek 2001.

$$P(\{\langle A_1^m, A_2^n, B_3^r, B_4^t\rangle \mid m,t \in \{-,+\}\}) =$$

$$\sum_{\lambda \in \Lambda} \mu_{13\Lambda}(\underline{a_1 b_3}, \lambda) \times \mu_{13\Lambda}(\underline{a_1 b_3}, \Lambda \mid \underline{a_1 b_3}, \lambda) \times \mu_{13\Lambda}(\underline{a_1 B_3^r}, \Lambda \mid \underline{a_1 b_3}, \lambda) \times$$

$$\times \mu_{24\Lambda}(\underline{A_2^n b_4}, \Lambda \mid \underline{a_2 b_4}, \lambda) \times \mu_{24\Lambda}(\underline{a_2 b_4}, \Lambda \mid \underline{a_2 b_4}, \lambda) =$$

$$\sum_{\lambda \in \Lambda} \mu_{13\Lambda}(\underline{a_1 b_3}, \lambda) \times 1 \times \mu_{13\Lambda}(\underline{a_1 B_3^r}, \Lambda \mid \underline{a_1 b_3}, \lambda) \times \mu_{24\Lambda}(\underline{A_2^n b_4}, \Lambda \mid \underline{a_2 b_4}, \lambda) \times 1 \overset{\text{NOCONS}}{=}$$

$$\sum_{\lambda \in \Lambda} \mu_{23\Lambda}(\underline{a_2 b_3}, \lambda) \times \mu_{13\Lambda}(\underline{a_1 B_3^r}, \Lambda \mid \underline{a_1 b_3}, \lambda) \times \mu_{24\Lambda}(\underline{A_2^n b_4}, \Lambda \mid \underline{a_2 b_4}, \lambda) \overset{\text{2xPI}}{=}$$

$$\sum_{\lambda \in \Lambda} \mu_{23\Lambda}(\underline{a_2 b_3}, \lambda) \times \mu_{23\Lambda}(\underline{a_1 B_3^r}, \Lambda \mid \underline{a_2 b_3}, \lambda) \times \mu_{23\Lambda}(\underline{A_2^n b_3}, \Lambda \mid \underline{a_2 b_3}, \lambda) \overset{\text{OI}}{=}$$

$$\sum_{\lambda \in \Lambda} \mu_{23\Lambda}(\underline{a_2 b_3}, \lambda) \times \mu_{23\Lambda}(\underline{A_2^n \cap B_3^r}, \Lambda \mid \underline{a_2 b_3}, \lambda) =$$

$$\mu_{23\Lambda}(A_2^n \cap B_3^r, \Lambda) = p_{23}(A_2^n \cap B_3^r) = \langle \psi \mid P_2^n P_3^r \psi \rangle$$

$$(6.15)$$

Finally, note that

$$P(\{\langle A_1^m, A_2^n, B_3^+, B_4^t\rangle \mid m,t \in \{-,+\}\})$$
$$+ P(\{\langle A_1^m, A_2^n, B_3^-, B_4^t\rangle \mid m,t \in \{-,+\}\}) = \qquad (6.16)$$
$$P(\{\langle A_1^m, A_2^n, B_3^r, B_4^t\rangle \mid m,r,t \in \{-,+\}\}) = p_{23}(A_2^n \cap b_3) = \langle \psi \mid P_2^n \psi \rangle$$

and analogously,

$$P(\{\langle A_1^m, A_2^n, B_3^r, B_4^t\rangle \mid m,n,t \in \{-,+\}\}) = p_{23}(a_2 \cap B_3^r) = \langle \psi \mid P_3^r \psi \rangle \qquad (6.17)$$

Since P is defined on atoms of $\mathcal{F}$, it is additive by the definition. Equations (6.15), (6.16), and (6.17) together with the definition of probability functions p_{ij} (Eq. (6.7)) show that P is normalized to unity, i.e., $P(\{\langle A_1^m, A_2^n, B_3^r, B_4^t\rangle \mid m,n,r,t \in \{-,+\}\}) = 1$. Thus, P is a probability measure, and $\langle \Omega, \mathcal{F}, P \rangle$ is a Kolmogorovian probability space, such that all the surface probabilities p_{ij} are identifiable with absolute probabilities in $\langle \Omega, \mathcal{F}, P \rangle$. □

The theorem thus shows that the three conditions OI, PI, NOCONS transform a hidden-state small Kolmogorovian representation of the Bell–Aspect experiment into a Kolmogorovian representation of this experiment (of Definition 2): the set $\{\langle A_1^m, A_2^n, B_3^r, B_4^t\rangle \mid n,r,t \in \{-,+\}\}$ corresponds to a single projector P_1^m present in the experiment, the set $\{\langle A_1^m, A_2^n, B_3^r, B_4^t\rangle \mid n,t \in \{-,+\}\}$ corresponds to a "double" projector $P_1^m P_3^r$ (and analogously for other projectors involved). Thus, whereas from a metaphysical perspective, the three conditions embody a claim about a causal underpinning of the Bell–Aspect experiment, their mathematical meaning is that they enforce pasting together of the four probability spaces (of the hidden-state small Kolmogorovian representation) into one probability space, constituting a Kolmogorovian representation of the Bell–Aspect experiment. By Pitowsky's theorem, the existence of the latter representation implies Bell's inequalities; and, to repeat, these inequalities are violated by quantum mechanics and most likely by Nature as well.

Given the mathematical meaning of the three conditions, one might want to draw a moral which perhaps suggests a further research program (this is a view of one co-author – TP, not shared by the other). Having seen what the three conditions do to a hidden-variable small Kolmogorovian representation, one should modify them so that the modified version do not enforce the pasting together of the small probability spaces. To please our causal intuitions, the modified OI, PI, and NOCONS should bear some relations to ideas underlying Reichenbach's common cause principle, locality, and no conspiracy. Hofer-Szabó's (2008) model with separate screener systems might be perhaps seen as a case in point (for an examination of the model, and some reservations towards it, see Wroński, Placek, and Godziszewski's paper in this volume). The mathematical meaning of the three conditions suggests also a modest program of testing models for the Bell–Aspect experiment, by asking whether the conditions of a given model imply the existence of Kolmogorovian representation of this experiment, or not.

6.6 Conclusions

According to the Pitowsky–Fine approach the violation of Bell's inequalities is the expression of the fact that quantum probabilities are not Kolmogorovian probabilities. However, as we learn from Szabó (1995, 2001), they should not be absolute Kolmogorovian probabilities in the first place: by virtue of their meaning, quantum probabilities are *conditional* and they reside in a large Kolmogorovian probability space that describes the measurement events of a real-world experiment.

There is a sense however in which Szabó's large space is not appropriate. Its construction essentially relies on an "excess structure" that does not seem to be inherent to the quantum phenomena. This excess structure corresponds to the probabilities of the different measurement settings that depends on the experimenter's decisions. We examined a simple method that factors out this excess structure to arrive at a collection of small Kolmogorovian probability spaces in which quantum probabilities are represented as *absolute* probabilities.

The idea that correlations between events residing in these small probability spaces call for a common causal explanation motivates a search for a hidden variable model of the experiment. We characterized such hidden variable models in terms of the small probability space representation of quantum probabilities. We showed that the existence of such a model allows us to paste together the small probability spaces representing the quantum probabilities in question into one single Kolmogorovian probability space that constitutes a Kolmogorovian representation of those quantum probabilities (in the original sense of Pitowsky–Fine).

In conclusion, one can phrase the duality of the two approaches to Bell's theorem this way: while the *physical* meaning of the violation of Bell's inequalities is the impossibility of certain causal explanations of the quantum phenomenon in question (respecting some physical/metaphysical principles about causation embodied by OI, PI, and NOCONS), the *mathematical* meaning of these principles is that they produce a Kolmogorovian probability representation of the phenomenon, which cannot exist (by Pitowski's theorem and the violation of Bell's inequalities).

Acknowledgements This work has been supported by the Hungarian Scientific Research Fund, OTKA K-115593 and by the Bilateral Mobility Grant of the Hungarian and Polish Academies of Sciences, NM-104/2014.

References

Accardi, L. 1984. The probabilistic roots of the quantum mechanical paradoxes. In *The wave-particle dualism*, ed. S. Diner et al. Dordrecht: Reidel.

Bell, J.S. 1964. On the Einsten–Podolsky–Rosen paradox. *Physics* 1: 195. Reprinted in Bell, J.S. 2004. *Speakable and unspeakable in quantum mechanics*. Cambridge: Cambridge University Press.

Bell, J.S. 1976. The theory of local beables. *Epistemological Letters*, March 1976. Reprinted in Bell, J.S. 2004. *Speakable and unspeakable in quantum mechanics*. Cambridge: Cambridge University Press.

Belnap, N., and L.E. Szabó. 1996. Branching space time analysis of the GHZ theorem. *Foundations of Physics* 26: 989.

Billingsley, P. 1995. *Probability and measure*, 3rd ed. New York: John Wiley & Sons, Inc.

Butterfield, J. 1992. Bell's theorem: What it takes. *British Journal for the Philosophy of Science* 43: 41–83.

Clauser, J.F., and A. Shimony. 1978. Bell's theorem: Experimental tests and implications. *Reports on Progress in Physics* 41: 1881.

Feintzeig, B. 2015. Hidden variables and incompatible observables in quantum mechanics. *The British Journal for the Philosophy of Science* 66(4): 905–927.

Fine, A. 1982a. Joint distributions, quantum correlations, and commuting observables. *Journal of Mathematical Physics* 23: 1306.

Fine, A. 1982b. Hidden variables, joint probability, and the Bell inequalities. *Physical Review Letters* 48: 291.

Gömöri, M., and G. Hofer-Szabó. 2017. On the meaning of EPR's Criterion of Reality. (In preparation).

Hofer-Szabó, G. 2008. Separate-versus common-common-cause-type derivations of the Bell inequalities. *Synthese* 163(2): 199–215.

Hofer-Szabó, G., M. Rédei, and L.E. Szabó. 2013. *The principle of the common cause*. Cambridge: Cambridge University Press.

Khrennikov, A.Y. 2010. *Ubiquitous quantum structure: From psychology to finance*. Berlin: Springer.

Kolmogorov, A.N. 1933. *Grundbegriffe der Wahrscheinlichkeitsrechnung*. Berlin: Julius Springer. English translation: Kolmogorov, A.N. 1956. *Foundations of the theory of probability*, 2nd ed. New York: Chelsea.

Müller, T., and T. Placek. 2001. Against a minimalist reading of Bell's theorem: Lessons from fine. *Synthese* 128(3): 343.

Pitowsky, I. 1989. *Quantum probability – Quantum logic*. Berlin: Springer.

Svetlichny, G., M. Redhead, H. Brown, and J. Butterfield. 1988. Do the Bell inequalities require the existence of joint probability distributions? *Philosophy of Science* 55(3): 387.

Szabó, L.E. 1995. Is quantum mechanics compatible with a deterministic universe? Two interpretations of o quantum probabilities. *Foundations of Physics Letters* 8: 417.

Szabó, L.E. 2001. Critical reflections on quantum probability theory. In *John von Neumann and the foundations of quantum physics*, ed. M. Rédei and M. Stoeltzner. Dordrecht: Kluwer.

Chapter 7
Reichenbachian Common Cause Systems of Size 3 in General Probability Theories

Yuichiro Kitajima

Abstract Reichenbach defined a common cause which explains a correlation between two events if either one does not cause the other. Its intuitive idea is that the statistical ensemble can be divided into two disjoint parts so that the correlation disappears in both of the resulting subensembles if there is no causal connection between these correlated events. These subensembles can be regarded as common causes. Hofer-Szabó and Rédei (Int J Theor Phys 43(7–8):1819–1826, 2004) generalized a Reichenbachian common cause, and called it a Reichenbachian common cause system. In the case of Reichenbachian common cause systems the statistical ensemble is divided more than two, while it is divided into two parts in the case of Reichenbachian common causes. The number of these subensembles is called the size of this system. In the present paper, we examine Reichenbachian common cause systems in general probability theories which include classical probability theories and quantum probability theories. It is shown that there is no Reichenbachian common cause system for any correlation between two events which are not logical independent, and that a general probability theory which is represented by an atomless orthomodular lattice with a faithful σ-additive probability measure contains Reichenbachian common cause systems of size 3 for any correlation between two compatible and logical independent events. Moreover, we discuss a relation between Bell's inequality and Reichenbachian common cause systems, and point out that this violation of Bell's inequality can be compatible with a Reichenbachian common cause system although it contradicts a 'common' common cause system.

Y. Kitajima (✉)
College of Industrial Technology, Nihon University, 2-11-1 Shin-ei, Narashino, Chiba 275-8576, Japan
e-mail: kitajima.yuichirou@nihon-u.ac.jp

© Springer International Publishing AG 2017
G. Hofer-Szabó, L. Wroński (eds.), *Making it Formally Explicit*, European Studies in Philosophy of Science 6, DOI 10.1007/978-3-319-55486-0_7

7.1 Introduction

Reichenbach (1956) defined a common cause which explain correlations between two events if either one does not cause the other. Its intuitive idea is that the statistical ensemble can be divided into two disjoint parts so that the correlation disappears in both of the resulting subensembles. These subensembles are regarded as common causes. Reichenbachain common cause principle states that a correlation between two events is either due to a direct causal link between the correlated events, or there is a Reichenbachian common cause that explains the correlation.

Gyenis and Rédei (2004) introduced the notion of common cause closedness, which means that there always exists a Reichenbachian common cause for any correlation between two events if either one does not cause the other. In a common cause closed probability measure, Reichenbachian common cause principle holds. When a probability measure space is not common cause closed, there are two strategies to save this principle. One strategies is to think that such a probability measure space is not rich enough to contain a common cause, and it can be embedded into a larger one which contains a common cause of the correlation. Such a larger probability space is called common cause complete (Hofer-Szabó et al. 2013, Chapter 3).

Another strategy is to suspect that the correlation is not due to two subensembles but the result of more than two subensembles. To formulate the latter idea, Hofer-Szabó and Rédei (2004, 2006) generalized a Reichenbachian common cause, and called it a Reichenbachian common cause system. The statistical ensemble is divided more than two in the case of Reichenbachian common cause systems while it is divided into two parts in the case of Reichenbachian common causes. The number of these subensembles is called the size of this system. A corresponding notion of common cause closedness can be introduced; a probability measure space is called causally n-closed if it contains Reichenbachian common cause systems whose subensembles are n for any correlation between two causally independent events.

Common causes closedness can be defined in quantum probability theories as well as classical probability theories (Gyenis and Rédei 2014; Kitajima and Rédei 2015). By a quantum probability theories are meant orthomodular lattices of projections of a noncommutative von Neumann algebra with σ-additive probability measure on this lattice. Kitajima and Rédei (2015) characterized common cause closedness of quantum probability theories. According to this result, a quantum probability space is common cause closed if and only if it has at most one measure theoretic atom.

The notion of Reichenbachian common cause systems also can be formulated in quantum probability theories as well as common cause closedness, and a more general notion of causal n-closedness of quantum probability theories can be defined naturally. Recently Wroński and Marczyk (2014, Section 6) examine causal n-closedness in an orthomodular lattice which has atoms.

Reichenbachian common cause systems are defined for the correlation between two events if either one does not cause the other. So it is important whether a causal connection between the two events exists or not. In orthomodular lattices, no causal connection is represented by logical independence (Rédei 1995a,b, 1998). In the present paper, we examine Reichenbachian common cause systems in orthomodular lattices. Concretely speaking, the following two problems are investigated:

- Is logical independence necessary for the existence of a Reichenbachian common cause system?
- Is there a general probability theory which is causal n-closed for some natural number n?

It is shown that there is no Reichenbachian common cause system for any correlation between two events if they are not logical independent (Theorem 10), and that a general probability theory which is represented by an atomless orthomodular lattice with a faithful σ-additive probability measure contains Reichenbachian common cause systems of size 3 for any correlation between two causally independent events (Theorem 12). In other words, any correlation in this probability theory can be explained by a Reichenbachian common cause system of size 3.

A reason why an atomless orthomodular lattice is examined is that it can be applied to algebraic quantum field theory (AQFT). AQFT predicts many states which violate Bell's inequality between two events associated with space-like separated spacetime regions in Minkowski spacetime (Halvorson and Clifton 2000; Kitajima 2013; Landau 1987; Summers and Werner 1987a,b,c, 1988). Does the existence of a Reichenbachian common cause system contradict this violation of Bell's inequality? Bell's inequality holds if it is assumed that there is a local hidden variable. This local hidden variable can be regarded as a 'common' common cause. Thus, no 'common' common cause exists if Bell's inequality does not hold. Because a Reichbachian common cause is different from 'common' common cause, it can be compatible with the violation of Bell's inequality (Rédei 1997; Hofer-Szabó et al. 1999; Rédei and Summers 2002). After Corollary 13, we discuss a relation between Bell's inequality and Reichenbachian common cause systems, and point out that this violation of Bell's inequality can be compatible with a Reichenbachian common cause system although it contradicts a 'common' common cause system.

7.2 Reichenbachian Common Cause Systems

An orthocomplemented lattice $\mathcal{L}$ with lattice operations $\vee$, $\wedge$, and orthocomplementation $\perp$ is called orthomodular if, for any $A, B \in \mathcal{L}$ such that $A \leq B$,

$$B = A \vee (A^{\perp} \wedge B). \tag{7.1}$$

Throughout the paper $\mathcal{L}$ denotes an orthomodular lattice. An example of orthomodular lattices $\mathcal{L}$ is a Boolean algebra, which is distributive, i.e. if for any $A, B, C \in \mathcal{L}$

$$A \vee (B \wedge C) = (A \vee B) \wedge (A \vee C).$$

Other examples of orthomodular lattices are the lattices of projections of a von Neumann algebra. The lattices of projections of a von Neumann algebra is distributive if and only if the von Neumann algebra is commutative. Generally an orthomodular lattice is not necessarily distributive.

The two elements $A, B \in \mathcal{L}$ be called compatible if

$$A = (A \wedge B) \vee (A \wedge B^\perp). \tag{7.2}$$

Equation (7.2) holds if and only if

$$B = (B \wedge A) \vee (B \wedge A^\perp). \tag{7.3}$$

In other words, the compatibility relation is symmetric (Kalmbach 1983, Theorem 3.2). If $A \leq B$, then A and B are compatible because $B = A \vee (A^\perp \wedge B) = (B \wedge A) \vee (B \wedge A^\perp)$.

In the present paper, it is assumed that orthomodular lattices are bounded: they have a smallest and a largest element denoted by 0 and 1, respectively. If for every countable subset S of $\mathcal{L}$, the join and the meet of all elements in S exist, then $\mathcal{L}$ is called a σ-complete orthomodular lattice. An orthomodular lattice is called atomless if, for any nonzero element $A \in \mathcal{L}$, there exists $B \in \mathcal{L}$ such that $0 < B < A$. For example, the orthomodular lattices of all projections on a Hilbert space is not atomless because one-dimensional projections are atoms in this case. On the other hand, the orthomodular lattices of projections of type II or type III von Neumann algebras are atomless orthomodular lattices. It is well known that typical local algebras in algebraic quantum field theory are of type III (Haag 1996, Section V.6), that is, the projection lattice of these algebras are atomless.

Let $\mathcal{L}$ be a σ-complete orthomodular lattice. Elements A and B in $\mathcal{L}$ are called mutually orthogonal if $A \leq B^\perp$. The map $\phi : \mathcal{L} \to [0, 1]$ is called a probability measure on $\mathcal{L}$ if $\phi(1) = 1$ and $\phi(A \vee B) = \phi(A) + \phi(B)$ for any mutually orthogonal elements A and B. A probability measure ϕ is called a σ-additive probability measure on $\mathcal{L}$ if for any countable, mutually orthogonal elements $\{A_i | i \in \mathbb{N}\}$,

$$\phi(\vee_{i \in \mathbb{N}} A_i) = \sum_{i \in \mathbb{N}} \phi(A_i).$$

If $\phi(A) = 0$ implies $A = 0$ for any $A \in \mathcal{L}$, then ϕ is called faithful.

Let $\mathcal{L}$ be an orthomodular lattice, ϕ be a probability measure on $\mathcal{L}$. We call $(\mathcal{L}, \phi)$ a nonclassical probability space. If $\mathcal{L}$ is a Boolean algebra, $(\mathcal{L}, \phi)$ is called a classical probability space. Thus nonclassical probability measure space is a more general notion than a classical one. An example of this space is a quantum probability space.

Definition 1 Let $\mathcal{L}$ be an orthomodular lattice, and let ϕ be a probability measure on $\mathcal{L}$. If A and B in $\mathcal{L}$ are compatible and

$$\phi(A \wedge B) > \phi(A)\phi(B),$$

then A and B are called correlated with respect to ϕ.

The following lemmas are needed in this paper.

Lemma 2 *Let $\mathcal{L}$ be an orthomodular lattice, and let ϕ be a probability measure on $\mathcal{L}$. Let A, B, and C in $\mathcal{L}$ be mutually compatible elements in $\mathcal{L}$. Then*

$$\phi(A \wedge C)\phi(B \wedge C) \geq \phi((A \wedge B) \wedge C)\phi((A \vee B) \wedge C)$$

Proof By Kalmbach (1983, p.25), the sublattice generated by A, B, and C is distributive. Thus

$$\phi(A \wedge C)\phi(B \wedge C)$$
$$= \phi(A \wedge C)(\phi(B \wedge C) - \phi(A \wedge B \wedge C)) + \phi(A \wedge C)\phi(A \wedge B \wedge C)$$
$$\geq \phi(A \wedge B \wedge C)(\phi(B \wedge C) - \phi(A \wedge B \wedge C)) + \phi(A \wedge C)\phi(A \wedge B \wedge C)$$
$$= \phi((A \wedge B) \wedge C)(\phi(A \wedge C) + \phi(B \wedge C) - \phi(A \wedge B \wedge C))$$
$$= \phi((A \wedge B) \wedge C)\phi((A \wedge C) \vee ((B \wedge C) \wedge (A \wedge B \wedge C)^{\perp}))$$
$$= \phi((A \wedge B) \wedge C)\phi((A \vee B) \wedge C).$$

$\square$

Lemma 3 *Let $\mathcal{L}$ be an orthomodular lattice, let ϕ be a probability measure on $\mathcal{L}$, and let A and B be elements in $\mathcal{L}$ such that $\phi(A \wedge B) > \phi(A)\phi(B)$. Then $\phi(A \vee B) < 1$ and $\phi(A \wedge B) > 0$.*

Proof Let A and B be mutually compatible elements in $\mathcal{L}$ such that

$$\phi(A \wedge B) > \phi(A)\phi(B). \tag{7.4}$$

Then $\phi(A \wedge B) > \phi(A)\phi(B) \geq 0$. By Lemma 2

$$\phi(A)\phi(B) \geq \phi(A \wedge B)\phi(A \vee B) \tag{7.5}$$

Equations (7.4) and (7.5) imply

$$\phi(A \vee B) < 1. \tag{7.6}$$

$\square$

Lemma 4 (Kitajima 2008, Lemma 3.6; Gyenis and Rédei 2014, Proposition 3.5)
Let $\mathcal{L}$ be a σ-complete atomless orthomodular lattice, ϕ a σ-additive probability measure on $\mathcal{L}$ and A an element in $\mathcal{L}$ such that $0 < \phi(A)$. For any real number x such that $0 < x < \phi(A)$, there is an element $X \in \mathcal{L}$ such that $\phi(X) = x$ and $0 < X < A$.

Reichenbachian common causes in a general probability theory are defined as follows.

Definition 5 (Hofer-Szabó et al. 2013, Definition 6.1) Let $\mathcal{L}$ be an orthomodular lattice, and let ϕ be a probability measure on $\mathcal{L}$. If A and B in $\mathcal{L}$ are correlated, then $C \in \mathcal{L}$ is called a common cause for the correlation if C is compatible with both A and B, and the following conditions hold:

$$\frac{\phi(A \wedge C)}{\phi(C)} \frac{\phi(B \wedge C)}{\phi(C)} = \frac{\phi(A \wedge B \wedge C)}{\phi(C)} \tag{7.7}$$

$$\frac{\phi(A \wedge C^{\perp})}{\phi(C^{\perp})} \frac{\phi(B \wedge C^{\perp})}{\phi(C^{\perp})} = \frac{\phi(A \wedge B \wedge C^{\perp})}{\phi(C^{\perp})} \tag{7.8}$$

$$\frac{\phi(A \wedge C)}{\phi(C)} > \frac{\phi(A \wedge C^{\perp})}{\phi(C^{\perp})} \tag{7.9}$$

$$\frac{\phi(B \wedge C)}{\phi(C)} > \frac{\phi(B \wedge C^{\perp})}{\phi(C^{\perp})} \tag{7.10}$$

The notion of Reichenbachian common causes can be generalized as follows.

Definition 6 Let $\mathcal{L}$ be an orthomodular lattice, and let ϕ be a probability measure on $\mathcal{L}$.

The set $\{C_j \in \mathcal{L} | j \in J\}$ is called a partition in $\mathcal{L}$ if $\vee_{j \in J} C_j = 1$, and C_i and C_j are orthogonal for $i \neq j$, where J is an index set.

If A and B in $\mathcal{L}$ are correlated, then a partition $\{C_j | j \in J\}$ is called a Reichenbachian common cause system for the correlation if C_j is compatible with both A and B for every $j \in J$, and the following conditions hold:

$$\frac{\phi(A \wedge B \wedge C_j)}{\phi(C_j)} = \frac{\phi(A \wedge C_j)}{\phi(C_j)} \frac{\phi(B \wedge C_j)}{\phi(C_j)} \tag{7.11}$$

for any $j \in J$, and

$$\left(\frac{\phi(A \wedge C_i)}{\phi(C_i)} - \frac{\phi(A \wedge C_j)}{\phi(C_j)} \right) \left(\frac{\phi(B \wedge C_i)}{\phi(C_i)} - \frac{\phi(B \wedge C_j)}{\phi(C_j)} \right) > 0 \tag{7.12}$$

for any mutually distinct elements $i, j \in J$. The cardinality of the index set J is called the size of the common cause system.

The following Lemma shows that A and B are correlated in ϕ if Equations (7.11) and (7.12) hold.

Lemma 7 (Hofer-Szabó and Rédei 2004, 2006) *Let $\mathcal{L}$ be an orthomodular lattice, let ϕ be a probability measure on $\mathcal{L}$, let $\{C_j | j \in J\}$ be a partition of $\mathcal{L}$ such that $\phi(C_j) > 0$ for any $j \in J$, and let A and B be mutually commuting elements in $\mathcal{L}$. If*

$$\frac{\phi(A \wedge B \wedge C_j)}{\phi(C_j)} = \frac{\phi(A \wedge C_j)}{\phi(C_j)} \frac{\phi(B \wedge C_j)}{\phi(C_j)},$$

for any $j \in J$, then

$$\phi(A \wedge B) - \phi(A)\phi(B) = \frac{1}{2} \sum_{i \neq j} \phi(C_i)\phi(C_j) \left(\frac{\phi(A \wedge C_i)}{\phi(C_i)} - \frac{\phi(A \wedge C_j)}{\phi(C_j)} \right)$$

$$\left(\frac{\phi(B \wedge C_i)}{\phi(C_i)} - \frac{\phi(B \wedge C_j)}{\phi(C_j)} \right)$$

Let A and B be compatible elements in an orthomodular lattice $\mathcal{L}$, and let $\mathcal{B}$ be the Boolean sub-lattice of $\mathcal{L}$ which contains A and B. If $A \wedge B^{\perp} = 0$, then $A = (A \wedge B) \vee (A \wedge (A \wedge B)^{\perp}) = (A \wedge B) \vee (A \wedge B^{\perp}) = A \wedge B$. It means that, for any truth-value assignment h of $\mathcal{B}$, $h(A) = 1$ entails $h(B) = 1$. If the truth-value of B is independent of that of A, $A \wedge B^{\perp}$ should not be 0. This is a motivation of logical independence

Definition 8 (Rédei 1995a,b, 1998) Let $\mathcal{L}$ be an orthomodular lattice, and let A and B be elements in $\mathcal{L}$. If $A \wedge B \neq 0$, $A^{\perp} \wedge B^{\perp} \neq 0$, $A \wedge B^{\perp} \neq 0$, and $A^{\perp} \wedge B \neq 0$, then it is said that A and B are logical independent.

Logical independence has the following property. This lemma is used in the proof of Theorem 12.

Lemma 9 *Let $\mathcal{L}$ be an orthomodular lattice, and let A and B be compatible elements in $\mathcal{L}$. Then the following conditions are equivalent.*

1. A and B are logical independent.
2. $A \vee B > A$, $A \vee B > B$, $A \vee B^{\perp} > A$, $A \vee B^{\perp} > B^{\perp}$

Proof $1 \Longrightarrow 2$ Let A and B be compatible and logical independent elements. Suppose $A = A \vee (A^{\perp} \wedge B)$. Then $A \geq A^{\perp} \wedge B$. Since A and B are compatible,

$$B = (B \wedge A) \vee (B \wedge A^{\perp}) \leq (B \wedge A) \vee A = A. \tag{7.13}$$

Then $A^{\perp} \wedge B \leq A^{\perp} \wedge A = 0$. It contradicts with logical independence between A and B.

The set of A, $A^\perp$, and B are distributive because A and B are compatible (Kalmbach 1983, p.25). Thus

$$A < A \vee (A^\perp \wedge B) = A \vee B \tag{7.14}$$

Similarly

$$B < B \vee (A \wedge B^\perp) = A \vee B, \tag{7.15}$$

$$A < A \vee (A^\perp \wedge B^\perp) = A \vee B^\perp, \tag{7.16}$$

$$B^\perp < (A \wedge B) \vee B^\perp = A \vee B^\perp. \tag{7.17}$$

$2 \Longrightarrow 1$ Let A and B be compatible elements in $\mathcal{L}$ which satisfies Condition 2. Then

$$A < A \vee B = A \vee (A^\perp \wedge B), \tag{7.18}$$

$$B < A \vee B = B \vee (A \wedge B^\perp), \tag{7.19}$$

$$A < A \vee B^\perp = A \vee (A^\perp \wedge B^\perp), \tag{7.20}$$

$$B^\perp < A \vee B^\perp = B^\perp \vee (A \wedge B). \tag{7.21}$$

Therefore, A and B are logical independent.

$\square$

The following theorem shows that logical independence plays an essential role in a Reichenbachian common cause system.

Theorem 10 *Let $\mathcal{L}$ be an orthomodular lattice, let ϕ be a probability measure on $\mathcal{L}$, let A and B be compatible elements in $\mathcal{L}$ such that $\phi(A \wedge B) > \phi(A)\phi(B)$, and let n be a natural number such that $n \geq 3$. If there exists a Reichenbachian common cause system of size n for the correlation between A and B, then A and B are logical independent.*

Proof Let A and B be compatible elements in $\mathcal{L}$ which are not logical independent and correlated in a probability measure ϕ on $\mathcal{L}$. By Lemma 3, $\phi(A \wedge B) > 0$ and $\phi(A^\perp \wedge B^\perp) > 0$. Thus

$$A \wedge B \neq 0, \quad A^\perp \wedge B^\perp \neq 0. \tag{7.22}$$

Since A and B are not logical independent, $A \wedge B^\perp = 0$ or $A^\perp \wedge B = 0$ by Equation (7.22). By symmetry we do not need to differentiate between $A^\perp \wedge B = 0$ and $A \wedge B^\perp = 0$, so we assume $A \wedge B^\perp = 0$.

Since A and B are compatible,

$$A = (A \wedge B) \vee (A \wedge B^\perp) = A \wedge B. \qquad (7.23)$$

Suppose that a Reichanbachian common cause system $\{C_1, C_2, \ldots, C_n\}$ for the correlation between A and B exists, where $n \geq 3$. By Equation (7.23)

$$\frac{\phi(A \wedge C_i)}{\phi(C_i)} \frac{\phi(B \wedge C_i)}{\phi(C_i)} = \frac{\phi(A \wedge B \wedge C_i)}{\phi(C_i)} = \frac{\phi(A \wedge C_i)}{\phi(C_i)}. \qquad (7.24)$$

Thus, for any $i \in \{1, 2, \ldots, n\}$, either $\phi(A \wedge C_i)/\phi(C_i) = 0$ or $\phi(B \wedge C_i)/\phi(C_i) = 1$ holds. Therefore there are mutually distinct natural numbers $i, j \in \{1, 2, \ldots, n\}$ such that either

$$\frac{\phi(A \wedge C_i)}{\phi(C_i)} = \frac{\phi(A \wedge C_j)}{\phi(C_j)} = 0 \quad \text{or} \quad \frac{\phi(B \wedge C_i)}{\phi(C_i)} = \frac{\phi(B \wedge C_j)}{\phi(C_j)} = 1 \qquad (7.25)$$

because $n \geq 3$. It implies

$$\left(\frac{\phi(A \wedge C_i)}{\phi(C_i)} - \frac{\phi(A \wedge C_j)}{\phi(C_j)} \right) \left(\frac{\phi(B \wedge C_i)}{\phi(C_i)} - \frac{\phi(B \wedge C_j)}{\phi(C_j)} \right) = 0. \qquad (7.26)$$

Thus $\{C_1, C_2, \ldots, C_n\}$ does not satisfy Equation (7.12) in Definition 6. Therefore no common cause system for the correlation between A and B exists. $\qquad \square$

According Theorem 10, if correlated events are not logical independent, there is no Reichenbachian common cause system for this correlation. Therefore logical independence is essential for Reichenbachian common cause systems.

A notion of causally n-closedness is obtained if the notion of common causes is replaced with the concept of common cause systems.

Definition 11 Let $\mathcal{L}$ be an orthomodular lattice, and let ϕ be a probability measure on $\mathcal{L}$. The probability space $(\mathcal{L}, \phi)$ is called causally n-closed with respect to logical independence if for any correlation between compatible and logical independent elements there exists a Reichenbachian common cause system of size n in $(\mathcal{L}, \phi)$.

In Theorem 12, we show that a general probability theory which is represented by an atomless orthomodular lattice contains Reichenbachian common cause systems of size 3 for any correlation between two logical independent events.

Theorem 12 Let $\mathcal{L}$ be a σ-complete atomless orthomodular lattice, and ϕ a faithful σ-additive probability measure on $\mathcal{L}$. Then the probability space $(\mathcal{L}, \phi)$ is causally 3-closed with respect to logical independence.

Proof Let A and B be mutually compatible and logical independent elements in $\mathcal{L}$ such that

$$\phi(A \wedge B) > \phi(A)\phi(B). \qquad (7.27)$$

By Lemma 2

$$\phi(A)\phi(B) \geq \phi(A \wedge B)\phi(A \vee B) \tag{7.28}$$

Equations (7.27) and (7.28) imply

$$0 < \phi(A \wedge B) - \phi(A)\phi(B) \leq \phi(A \wedge B)(1 - \phi(A \vee B)), \tag{7.29}$$

and

$$\phi(A \vee B) < 1 \tag{7.30}$$

by Lemma 3. Inequalities (7.29) and (7.30) entail

$$0 < \frac{\phi(A \wedge B) - \phi(A)\phi(B)}{1 - \phi(A \vee B)} \leq \phi(A \wedge B). \tag{7.31}$$

By Lemma 4 and Inequality (7.31) there exists an element C_1 such that

$$0 < C_1 < A \wedge B \tag{7.32}$$

and

$$0 < \phi(C_1) < \frac{\phi(A \wedge B) - \phi(A)\phi(B)}{1 - \phi(A \vee B)} \leq \phi(A \wedge B). \tag{7.33}$$

Since C_1 is compatible with $A \wedge B, A$, and B,

$$\phi(X \wedge C_1^{\perp}) = \phi(X) - \phi(X \wedge C_1) = \phi(X) - \phi(C_1) > 0, \tag{7.34}$$

where $X = A \wedge B, A, B$. By Inequality (7.32)

$$\frac{\phi(A \wedge C_1)}{\phi(C_1)} = \frac{\phi(B \wedge C_1)}{\phi(C_1)} = \frac{\phi(A \wedge B \wedge C_1)}{\phi(C_1)} = 1. \tag{7.35}$$

Inequality (7.33) and Equation (7.34) imply

$$\phi(C_1^{\perp})\phi(A \wedge B \wedge C_1^{\perp}) - \phi(A \wedge C_1^{\perp})\phi(B \wedge C_1^{\perp})$$
$$= (1 - \phi(C_1))(\phi(A \wedge B) - \phi(C_1)) - (\phi(A) - \phi(C_1))(\phi(B) - \phi(C_1))$$
$$(\because \text{Equation (7.34)})$$
$$= \phi(A \wedge B) - \phi(A)\phi(B) - \phi(C_1)(1 - \phi(A) - \phi(B) + \phi(A \wedge B))$$
$$= \phi(A \wedge B) - \phi(A)\phi(B) - \phi(C_1)(1 - \phi(A \vee B))$$
$$> 0 \quad (\because \text{Inequality (7.33)})$$

$$\tag{7.36}$$

Since C_1 is compatible with both A and B,

$$\phi(A \wedge C_1^\perp)\phi(B \wedge C_1^\perp) \geq \phi((A \wedge B) \wedge C_1^\perp)\phi((A \vee B) \wedge C_1^\perp), \qquad (7.37)$$

by Lemma 2.

By Inequality (7.32) $C_1^\perp > (A \wedge B)^\perp \geq (A \vee B)^\perp$. Thus $C_1^\perp$ and $(A \vee B)^\perp$ are compatible. Therefore

$$\phi(C_1^\perp) = \phi(C_1^\perp \wedge (A \vee B)^\perp) + \phi(C_1^\perp \wedge (A \vee B)) = \phi((A \vee B)^\perp) + \phi((A \vee B) \wedge C_1^\perp).$$
$$(7.38)$$

Inequalities (7.34) and (7.36) entail

$$0 < \phi(C_1^\perp) - \frac{\phi(A \wedge C_1^\perp)\phi(B \wedge C_1^\perp)}{\phi(A \wedge B \wedge C_1^\perp)}$$

$$\leq \phi(C_1^\perp) - \phi((A \vee B) \wedge C_1^\perp) \quad (\because \text{Inequality (7.37)}) \qquad (7.39)$$

$$= \phi((A \vee B)^\perp) \quad (\because \text{Equation (7.38)})$$

$$= \phi(A^\perp \wedge B^\perp).$$

By Lemma 4 there exists an element $C_2 \in \mathcal{L}$ such that

$$C_2 \leq A^\perp \wedge B^\perp \qquad (7.40)$$

and

$$\phi(C_2) = \phi(C_1^\perp) - \frac{\phi(A \wedge C_1^\perp)\phi(B \wedge C_1^\perp)}{\phi(A \wedge B \wedge C_1^\perp)} > 0. \qquad (7.41)$$

By Inequality (7.40)

$$\frac{\phi(A \wedge C_2)}{\phi(C_2)} = \frac{\phi(B \wedge C_2)}{\phi(C_2)} = \frac{\phi(A \wedge B \wedge C_2)}{\phi(C_2)} = 0. \qquad (7.42)$$

Let $C_3 := C_1^\perp \wedge C_2^\perp$. Then C_1, C_2, and C_3 are mutually orthogonal elements such that $C_1 \vee C_2 \vee C_3 = 1$. Thus $\{C_1, C_2, C_3\}$ is a partition in $\mathcal{L}$. Since $(A \wedge B) \vee (A \wedge B^\perp) \vee (A^\perp \wedge B) \vee (A^\perp \wedge B^\perp) = 1$ by Kalmbach (1983, p.26),

$$\phi(A \wedge B) + \phi(A \wedge B^\perp) + \phi(A^\perp \wedge B) + \phi(A^\perp \wedge B^\perp) = 1. \qquad (7.43)$$

Thus

$$\phi(C_3) = 1 - \phi(C_1) - \phi(C_2)$$

$$> 1 - \phi(A \wedge B) - \phi(A^\perp \wedge B^\perp)$$

$$(\because \text{Equation (7.41), and Inequalities (7.33) and (7.39))}$$

$$= \phi(A \wedge B^\perp) + \phi(A^\perp \wedge B) \quad (\because \text{Equation (7.43))}$$

$$\geq 0.$$

$$(7.44)$$

By Equation (7.41)

$$\frac{\phi(A \wedge C_3)}{\phi(C_3)} \frac{\phi(B \wedge C_3)}{\phi(C_3)} = \frac{\phi(A \wedge B \wedge C_3)}{\phi(C_3)} \tag{7.45}$$

since

$$\phi(C_3) = 1 - \phi(C_1) - \phi(C_2) = \phi(C_1^\perp) - \phi(C_2) \tag{7.46}$$

and

$$X \wedge C_3 = X \wedge C_1^\perp \wedge C_2^\perp = X \wedge C_1^\perp \tag{7.47}$$

where $X = A, B, A \wedge B$.

By Lemma 9

$$\phi(A \vee B) > \phi(A), \quad \phi(A \vee B) > \phi(B) \tag{7.48}$$

since ϕ is faithful, and A and B are compatible and logical independent.

Equation (7.47) and Inequality (7.33) imply

$$\phi(A \wedge C_3) = \phi(A \wedge C_1^\perp) = \phi(A) - \phi(C_1) > 0$$

$$\phi(B \wedge C_3) = \phi(B \wedge C_1^\perp) = \phi(B) - \phi(C_1) > 0, \tag{7.49}$$

and Inequalities (7.39) and (7.48), and Equation (7.41) entail

$$\phi(C_2) \leq \phi(A^\perp \wedge B^\perp) = 1 - \phi(A \vee B) < 1 - \phi(A)$$

$$\phi(C_2) \leq \phi(A^\perp \wedge B^\perp) = 1 - \phi(A \vee B) < 1 - \phi(B). \tag{7.50}$$

Thus

$$0 < \frac{\phi(A \wedge C_3)}{\phi(C_3)} = \frac{\phi(A) - \phi(C_1)}{1 - \phi(C_2) - \phi(C_1)} < \frac{1 - \phi(C_2) - \phi(C_1)}{1 - \phi(C_2) - \phi(C_1)} = 1$$

$$0 < \frac{\phi(B \wedge C_3)}{\phi(C_3)} = \frac{\phi(B) - \phi(C_1)}{1 - \phi(C_2) - \phi(C_1)} < \frac{1 - \phi(C_2) - \phi(C_1)}{1 - \phi(C_2) - \phi(C_1)} = 1. \tag{7.51}$$

By Equations (7.35) and (7.42), and Inequality (7.51)

$$\left(\frac{\phi(A \wedge C_i)}{\phi(C_i)} - \frac{\phi(A \wedge C_j)}{\phi(C_j)}\right)\left(\frac{\phi(B \wedge C_i)}{\phi(C_i)} - \frac{\phi(B \wedge C_j)}{\phi(C_j)}\right) > 0 \quad (i \neq j). \quad (7.52)$$

□

By Proposition 10 and Theorem 12, we get the following corollary.

Corollary 13 *Let $\mathcal{L}$ be a σ-complete atomless orthomodular lattice, let ϕ be a faithful σ-additive probability measure on $\mathcal{L}$, and let A and B be compatible elements in $\mathcal{L}$ such that $\phi(A \wedge B) > \phi(A)\phi(B)$. There exists a Reichenbachian common cause system of size 3 for the correlation between A and B if and only if A and B are logical independent.*

This corollary shows that the existence of a Reichenbachian common cause system of size 3 for the correlation between A and B is equivalent to logical independence of A and B in the case of a general probability theory which is represented by a σ-complete atomless orthomodular lattice with a faithful σ-additive probability measure.

Theorem 12 and Corollary 13 can apply to AQFT because typical local algebras in algebraic quantum field theory are type III von Neumann algebras and lattices of projections of type II or type III von Neumann algebras is atomless orthomodular lattices. In algebraic quantum field theory, each bounded open region $\mathcal{O}$ in the Minkowski space is associated with a von Neumann algebra $\mathfrak{N}(\mathcal{O})$. Such a von Neumann algebra is called a local algebra. We say that bounded open regions $\mathcal{O}_1$ and $\mathcal{O}_2$ are strictly space-like separated if there is a neighborhood V of the origin of the Minkowski space such that $\mathcal{O}_1 + V$ and $\mathcal{O}_2$ are space-like separated.

If $\mathcal{O}_1$ and $\mathcal{O}_2$ are strictly space-like separated bounded open regions in the Minkowski space, local algebras $\mathfrak{N}(\mathcal{O}_1)$ and $\mathfrak{N}(\mathcal{O}_2)$ are logical independent under usual axioms (Baumgärtel 1995, Theorem 1.12.3). Thus there exists a Reichenbachian common cause system of size 3 for any correlation between $\mathfrak{N}(\mathcal{O}_1)$ and $\mathfrak{N}(\mathcal{O}_2)$ by Corollary 13. On the other hand, it is known that there is a normal state on $\mathfrak{N}(\mathcal{O}_1) \vee \mathfrak{N}(\mathcal{O}_2)$ which violates Bell's inequality (Halvorson and Clifton 2000; Kitajima 2013; Landau 1987; Summers and Werner 1987a,b,c, 1988). In other words, there exist a normal state ϕ on $\mathfrak{N}(\mathcal{O}_1) \vee \mathfrak{N}(\mathcal{O}_2)$, and self-adjoint contractions $A_1, B_1 \in \mathfrak{N}(\mathcal{O}_1)$ and $A_2, B_2 \in \mathfrak{N}(\mathcal{O}_2)$ such that

$$\frac{1}{2}|\phi(A_1A_2 + A_1B_2 + B_1A_2 - B_1B_2)| > 1. \quad (7.53)$$

Suppose that there is a Reichenbachian 'common' common cause system $\{C_1, \ldots, C_n\}$ for all four pairs (A_1, A_2), (A_1, B_2), (B_1, A_2), and (B_1, B_2). In other words,

$$\frac{\phi(A_1A_2C_j)}{\phi(C_j)} = \frac{\phi(A_1C_j)}{\phi(C_j)}\frac{\phi(A_2C_j)}{\phi(C_j)},$$

$$\frac{\phi(A_1B_2C_j)}{\phi(C_j)} = \frac{\phi(A_1C_j)}{\phi(C_j)}\frac{\phi(B_2C_j)}{\phi(C_j)},$$

$$\frac{\phi(B_1A_2C_j)}{\phi(C_j)} = \frac{\phi(B_1C_j)}{\phi(C_j)}\frac{\phi(A_2C_j)}{\phi(C_j)},$$

$$\frac{\phi(B_1B_2C_j)}{\phi(C_j)} = \frac{\phi(B_1C_j)}{\phi(C_j)}\frac{\phi(B_2C_j)}{\phi(C_j)},$$

$$[A_1, C_j] = [A_2, C_j] = [B_1, C_j] = [B_2, C_j] = 0,$$

$$\sum_{i=1}^{n} C_i = I,$$

(7.54)

for any $j \in \{1, \ldots, n\}$.

Then

$$\frac{1}{2}\left|\phi(A_1A_2 + A_1B_2 + B_1A_2 - B_1B_2)\right|$$

$$= \frac{1}{2}\left|\phi\left((A_1A_2 + A_1B_2 + B_1A_2 - B_1B_2)\left(\sum_{i=1}^{n} C_i\right)\right)\right|$$

$$= \frac{1}{2}\left|\sum_{i=1}^{n} \phi\left((A_1A_2 + A_1B_2 + B_1A_2 - B_1B_2)C_i\right)\right|$$

$$= \frac{1}{2}\left|\sum_{i=1}^{n} (\phi(A_1A_2C_i) + \phi(A_1B_2C_i) + \phi(B_1A_2C_i) - \phi(B_1B_2C_i))\right|$$

$$= \frac{1}{2}\left|\sum_{i=1}^{n} \left(\frac{\phi(A_1A_2C_i)}{\phi(C_i)}\phi(C_i) + \frac{\phi(A_1B_2C_i)}{\phi(C_i)}\phi(C_i) + \frac{\phi(B_1A_2C_i)}{\phi(C_i)}\phi(C_i) \right.\right.$$

$$\left.\left. - \frac{\phi(B_1B_2C_i)}{\phi(C_i)}\phi(C_i)\right)\right|$$

$$\leq \frac{1}{2}\sum_{i=1}^{n} \left(\left|\frac{\phi(A_1C_i)}{\phi(C_i)}\frac{\phi(A_2C_i)}{\phi(C_i)} + \frac{\phi(A_1C_i)}{\phi(C_i)}\frac{\phi(B_2C_i)}{\phi(C_i)} + \frac{\phi(B_1C_i)}{\phi(C_i)}\frac{\phi(A_2C_i)}{\phi(C_i)}\right.\right.$$

$$\left.\left. - \frac{\phi(B_1C_i)}{\phi(C_i)}\frac{\phi(B_2C_i)}{\phi(C_i)}\right|\phi(C_i)\right)$$

$$\leq \frac{1}{2} \cdot 2\left(\sum_{i=1}^{n} \phi(C_i)\right)$$

$$= 1$$

(7.55)

by Equations (7.54). It contradicts Inequality (7.53). Thus, there is no Reichenbachian 'common' common cause system $\{C_1, \ldots, C_n\}$ for all four pairs (A_1, A_2), (A_1, B_2), (B_1, A_2), and (B_1, B_2) because Bell's inequality does not hold in ϕ. It means that the existence of Reichenbachian 'common' cause system contradicts the violation of Bell's inequality (Rédei 1997; Hofer-Szabó et al. 1999; Rédei and Summers 2002).

On the other hand, the definition of Reichenbachian common cause systems does not require that it is common for all four pairs (Definition 6). Thus, the existence of a Reichenbachian common cause system can be compatible with the violation of Bell's inequality in AQFT, and Theorem 12 and Corollary 13 show that there are Reichenbachian common cause systems of size 3 for any correlation between $\mathfrak{N}(\mathcal{O}_1)$ and $\mathfrak{N}(\mathcal{O}_2)$ in spite of the violation of Bell's inequality.

7.3 Conclusion

In the present paper, we examined Reichenbachian common cause systems in general probability theories which include classical probability theories and quantum probability theories. Especially, the following two problems were investigated:

- Is logical independence necessary for the existence of a Reichenbachian common cause system?
- Is there a general probability theory which is causal n-closed for some natural number n?

In Theorem 10, it was shown that there is no Reichenbachian common cause system for a correlation between two events if they are not logical independent. Therefore logical independence is necessary for the existence of a Reichenbachian common cause system. In Theorem 12 it was shown that a general probability theory which is represented by an atomless orthomodular lattice with a faithful σ-additive probability measure is causally 3-closed. It remains open, however, whether this lattice is causally n-closed for any natural number n.

Theorem 12 and Corollary 13 can apply to AQFT because typical local algebras in algebraic quantum field theory are type III von Neumann algebras and lattices of projections of type II or type III von Neumann algebras is atomless orthomodular lattices. There exists a Reichenbachian common cause system of size 3 for any correlation between two space-like separated regions by Corollary 13. On the other hand, AQFT predicts many states which violate Bell's inequality between two events associated with space-like separated spacetime regions in Minkowski spacetime (Halvorson and Clifton 2000; Kitajima 2013; Landau 1987; Summers and Werner 1987a,b,c, 1988). After Corollary 13, we pointed out that this violation of Bell's inequality can be compatible with a Reichenbachian common cause system although it contradicts a 'common' common cause system.

Rédei and Summers (2002, 2007) showed that a Reichenbachian common cause for any correlation exists in the union of the backward light cones of $\mathcal{O}_1$ and $\mathcal{O}_2$

in AQFT. By this result and Theorem 12, it can be shown that a Reichenbachian common cause system of size 3 for any correlation exists in the union of the backward light cones of $\mathcal{O}_1$ and $\mathcal{O}_2$. However, the problem about Reichenbachian commmon cause systems in AQFT is largely open.

Acknowledgements The author is supported by the JSPS KAKENHI No.15K01123 and No.23701009.

References

Baumgärtel, H. 1995. *Operatoralgebraic methods in quantum field theory*. Berlin: Akademie Verlag.

Gyenis, B., and M. Rédei. 2004. When can statistical theories be causally closed? *Foundations of Physics* 34(9): 1285–1303.

Gyenis, Z., and M. Rédei. 2014. Atomicity and causal completeness. *Erkenntnis* 79(3): 437–451.

Haag, R. 1996. *Local quantum physics: Fields, particles, algebras*. Berlin, Heidelberg, New York: Springer.

Halvorson, H., and R. Clifton. 2000. Generic Bell correlation between arbitrary local algebras in quantum field theory. *Journal of Mathematical Physics* 41(4): 1711–1717.

Hofer-Szabó, G., and M. Rédei. 2004. Reichenbachian common cause systems. *International Journal of Theoretical Physics* 43(7–8): 1819–1826.

Hofer-Szabó, G., and M. Rédei. 2006. Reichenbachian common cause systems of arbitrary finite size exist. *Foundations of Physics* 36(5): 745–756.

Hofer-Szabó, G., M. Rédei, and L.E. Szabo. 1999. On Reichenbach's common cause principle and Reichenbach's notion of common cause. *The British Journal for the Philosophy of Science* 50(3): 377–399.

Hofer-Szabó, G., M. Rédei, and L.E. Szabó. 2013. *The principle of the common cause*. Cambridge: Cambridge University Press.

Kalmbach, G. 1983. *Orthomodular lattices*. London: Academic Press.

Kitajima, Y. 2008. Reichenbach's common cause in an atomless and complete orthomodular lattice. *International Journal of Theoretical Physics* 47(2): 511–519.

Kitajima, Y. 2013. EPR states and Bell correlated states in algebraic quantum field theory. *Foundations of Physics* 43(10): 1182–1192.

Kitajima, Y., and M. Rédei. 2015. Characterizing common cause closedness of quantum probability theories. *Studies in History and Philosophy of Modern Physics* 52: 234–241.

Landau, L.J. 1987. On the violation of Bell's inequality in quantum theory. *Physics Letters A* 120(2): 54–56.

Rédei, M. 1995a. Logical independence in quantum logic. *Foundations of Physics* 25(3): 411–422.

Rédei, M. 1995b. Logically independent von Neumann lattices. *International Journal of Theoretical Physics* 34(8): 1711–1718.

Rédei, M. 1997. Reichenbach's common cause principle and quantum field theory. *Foundations of physics* 27(10): 1309–1321.

Rédei, M. 1998. *Quantum logic in algebraic approach*. Dordrecht: Springer.

Rédei, M., and S.J. Summers. 2002. Local primitive causality and the common cause principle in quantum field theory. *Foundations of Physics* 32(3): 335–355.

Rédei, M., and S.J. Summers. 2007. Remarks on causality in relativistic quantum field theory. *International Journal of Theoretical Physics* 46(8): 2053–2062.

Reichenbach, H. 1956. *The direction of time*. Berkeley: University of California Press.

Summers, S. J., and R. Werner. 1987a. Bell's inequalities and quantum field theory. I. general setting. *Journal of Mathematical Physics* 28(10): 2440–2447.

Summers, S. J., and R. Werner. 1987b. Bell's inequalities and quantum field theory. II. Bell's inequalities are maximally violated in the vacuum. *Journal of Mathematical Physics* 28(10): 2448–2456.

Summers, S. J., and R. Werner. 1987c. Maximal violation of Bell's inequalities is generic in quantum field theory. *Communications in Mathematical Physics* 110(2): 247–259.

Summers, S. J., and R. Werner. 1988. Maximal violation of Bell's inequalities for algebras of observables in tangent spacetime regions. *Annales de l'IHP Physique théorique* 49(2): 215–243.

Wroński, L., and M. Marczyk. 2014. A new notion of causal closedness. *Erkenntnis* 79(3): 453–478.

Chapter 8
On Time Order and Causal Order in the EPR Experiment

Iñaki San Pedro

Abstract The aim of this paper is to discuss the rejection of the so-called *Measurement Independence*—i.e. *No-conspiracy*—condition, in the context of causal explanations of EPR correlations, and survey some of its implications. In particular, I pay attention here to a specific way *Measurement Independence* is violated. It has to do with two assumptions about the presupposed causal order and space-time arrangement of the events involved in the EPR picture. The consequences are mostly, and more importantly, related to locality issues.

8.1 Introduction

Ever since Bas van Fraassen's influential "Charybdis of Realism ..." (van Fraassen 1982) it is a widespread opinion among philosophers of science that common cause accounts of the EPR violations are to be ruled out. This is because, as van Fraassen's paper shows, the idea of common cause (Reichenbach 1956) can be identified with that of hidden variable, as in Bell's theorem (Bell 1964). Thus, van Fraassen argues, postulating the existence of (screening-off) common cause events for the EPR correlations leads to the Bell inequalities, which are known to be empirically violated. One should then conclude, following the very consequences of Bell's theorem, that common cause explanations of EPR are not a possibility, i.e. that the hidden common cause variable that such accounts presuppose simply does not exist.

Such views attracted huge attention at the time and gave rise to a large literature on whether the idea of common cause is sensible and useful a notion to explain the EPR correlations. The general agreement being, as already pointed out, that

Financial support is gratefully acknowledged from BBVA Foundation, research project "Causality, Locality and Free Will in Quantum Mechanics", and Spanish Ministry of Economy, Industry and Competitiveness, project FFI2014-57064-P.

I. San Pedro (✉)
Department of Logic and Philosophy of Science, University of the Basque Country, UPV/EHU, Av. Tolosa 70, 28018 Donostia-San Sebastián, Spain
e-mail: inaki.sanpedro@ehu.eus

© Springer International Publishing AG 2017   147
G. Hofer-Szabó, L. Wroński (eds.), *Making it Formally Explicit*, European Studies in Philosophy of Science 6, DOI 10.1007/978-3-319-55486-0_8

EPR correlations cannot be accounted for in terms of common causes. Not that van Fraassen's derivation of the Bell inequalities was not contested at all. For instance, Hofer-Szabó et al. (2002) criticised the argument in "Charybdis of Realism ..." on the grounds that van Fraassen's definition of common cause does not exactly match that of Reichenbach's original proposal. In particular, van Fraassen's common causes are defined to be a rather restricted kind of events, which are required to screen-off two or more correlations at once. (These are so-called *common*-common causes, in contrast to Reichenbach's original *simple*, or *separate*, common causes.) Further refined versions of van Fraassen's original derivation have nevertheless reached similar conclusions than that in "Charybdis of Realism ...", reinforcing the idea that Reichenbachian common causes cannot account for the EPR correlations.[1]

Typical derivations of the Bell inequalities presuppose a common cause on to which several constraints and restrictions are set. Constraints on the postulated common causes are intended to reflect standard requirements of a generic physical system, including temporal order of causal relations or locality considerations. As a result, some version of Bell's *factorizability*—and therefore of a Bell-type inequality—is derived. The strength of such arguments relies thus on the plausibility of the conditions imposed on the common causes. There is, for instance, an extensive literature regarding the idea of locality, particularly concerning the intuitions leading to the concept of physical locality, the characterisation of the concept itself, its implications and whether it may be appropriately captured and characterised in terms of probabilistic relations.

Less attention has been paid to the requirement that the EPR experimenters do take free independent decisions at the moment of setting up the EPR apparatus for measurement. Roughly, this is usually taken to entail that the events representing the experimenters' decisions, and the foregoing corresponding free acts, be causally independent of the hidden variables. This is usually expressed by means of the so-called *No-conspiracy* condition—I shall in what follows refer to this condition, more neutrally, as *Measurement Independence*—, a probabilistic expression which is in some occasions taken to be necessary for free will.[2]

Rejecting *Measurement Independence*, however, is still an interesting option. Indeed, we might have good reasons for entertaining this possibility, as I already suggested in San Pedro (2013). These are mainly related to the different fashions in which *Measurement Independence* can be violated. The aim of this paper is thus to discuss and elaborate further some of the implications resulting form the rejection of *Measurement Independence*. In particular, the paper is concerned with a specific way the condition is violated, as a consequence of the rejection of two specific assumptions about the presupposed causal order and space-time arrangement of the events involved in the EPR picture.

---

[1] See, for instance, (Graßhoff et al. 2005) for a more recent example.

[2] Reference to 'free will' in this context is usually set aside in favour of more general stronger claims about 'world (or cosmic) conspiracies' instead. The exact relation between the requirement of *Measurement Independence*, 'free will' and 'world conspiracies' will be addressed more in detail in what follows.

The paper is divided in two parts. First, I shall review briefly the arguments for the requirement of *Measurement Independence* in the EPR context. Sections 8.2 and 8.3, in particular, provide an account of the general structure of the problem and the different arguments against *Measurement Independence* respectively, with special emphasis on the arguments as regards specific space-time and causal presuppositions behind it. In the second part of the paper I take on the implications of the actual violation of *Measurement Independence* as a consequence of the rejection of such space-time and/or causal assumptions. Here, the three resulting causal models initially hinted in San Pedro (2013) are discussed. The paper closes with some brief remarks on the issue.

8.2 Free Will, Conspiracy and *Measurement Independence*

Measurement Independence is the requirement that common causes C postulated to explain the EPR correlations be probabilistically independent from the corresponding measurement operations m_i performed on either wing of the experiment, i.e.

$$p(m_i|C) = p(m_i). \tag{8.1}$$

(I will write m_i for a generic measurement operation in an EPR experiment, with $i = L, R$ indicating that measurement is performed on the left and right wings of the experiment respectively. Similarly, in what follows, O_i, with $i = L, R$, will denote generic outcomes of the experiment.)

That Equation (8.1) must hold in any common cause explanation of the EPR correlations is often justified by the fact that EPR experimenters act freely to choose which specific measurements to perform each time. This requirement for free will in itself does not seem to be at all controversial. In particular, it seems desirable that any theory we propose that aims at a description of nature and that may include or refer to our (human) interaction with it, be consistent with the idea of free will—unless, of course, we discard the possibility of free agents from the very start. A more interesting matter concerns the issue as to how to represent appropriately the idea of free will within the theory, be it as a piece of mathematical formalism, as some set of background assumptions or presuppositions, etc. Addressing such issues however would take us far from our purpose here, since we are just concerned, at least in a first instance, with the more specific question whether there is indeed a relation between *Measurement Independence* and the idea of free will.

So does Equation (8.1) adequately represent a requirement related to the preservation of free will in the EPR context?

Many would claim already that this actually is not the right question to ask. For perhaps *Measurement Independence* has nothing todo with us humans having freedom of will, really, but rather with the idea that there be no (cosmic) conspiracies. As I will suggest later, however, these two claims are related, one being the generalisation of the other, i.e. claims about conspiracies are a generalisation of claims about the lack of free will. Let me then start with the arguments regarding the more specific requirement of free will.

So, once more, is this really so, i.e. is *Measurement Independence* really a requirement about free will? This question is very seldom addressed in the literature. In most derivations of the Bell inequalities the assumption of *Measurement Independence*—or for that matter *No-conspiracy*—is usually introduced rather uncritically and without proper justification. In particular, why is it that *Measurement Independence* guarantees the preservation of free will, or the lack of world conspiracies, is almost never addressed.

I have discussed the issue in some detail before in San Pedro (2013). The arguments there actually point to a conceptual independence between *Measurement Independence* and free will. We need not review such arguments in detail here but perhaps a brief sketch of them is in order—especially since the discussion that follows draws on one of these specific arguments.

In (San Pedro 2013) I point out in the first place that there are at least two ways to motivate a close relation between a statistical conditions such as *Measurement Independence* and the idea of free will. Namely, one may want to build an account of free will in terms of probabilistic relations from scratch—e.g. by defining acts of will in the first place and then providing a formalism which is able to accommodate dependence/independence between them. Alternatively, one may take the less ambitious option of identifying central features associated to free will which may have a more or less straightforward translation into probabilistic terms. Causation, or causal relations, seem like a good candidate if we are to pursue this later strategy. In fact, the notion of free will involves and presupposes a number of causal assumptions.

Thus, by paying a closer attention to those causal presuppositions behind the idea of free will which carry over to the formulation of *Measurement Independence*, we are in a position to address whether the two are indeed related or not.[3] In San Pedro (2013) I identify three such causal assumptions. I note that for *Measurement Independence* to represent some idea of free will one needs to assume (i) that there is a faithful connection between causal relations and statistical relations, i.e. *cause-statistics link* assumption, (ii) that the events involved have a precise fixed temporal arrangement, i.e. *time order* assumption, and (iii) that there are no causal influences *at all* between the postulated common causes and the events representing the corresponding settings of the experiment (and therefore between the common cause and the experimenters decisions), i.e. *no-cause* assumption (San Pedro 2013, pp. 92–94).

---

[3]I would even say that it is indeed the fact that these assumptions behind the idea of free will carry over to *Measurement Independence* what it is most often seen as justifying that the later stands for the former. And my point in San Pedro (2013) is precisely that such assumptions are not well grounded and can all be challenged.

All the above assumptions can be challenged on their own grounds to show that the putative link between the notion of free will and *Measurement Independence* is, if at all, weaker than initially supposed. Since in what follows I shall expand on the original argument concerning assumption (ii), i.e. *time order*, to the more general case of world conspiracies (see below), let me just briefly comment on assumptions (i) and (iii) above. Starting with the assumption that there is a faithful correspondence between causal relations and probabilistic statements it is clearly a strong idealisation. It is worth noting however that we need this assumption to be in place if our aim is to give causal explanations of correlations at all. Thus, we must assume the *cause-statistics link* even if it is not fully justified to do so if we attempt to provide a causal explanation of EPR correlations (San Pedro 2013, pp. 94–95). As regards assumption (iii) above, i.e. *no-cause*, it entails that the postulated common cause is either a deterministic common cause or at least a total cause of the measurement settings. In the face of it, this seems to be too strong an assumption (San Pedro 2013, pp. 99–100).

Despite the conclusion suggesting there is no conceptual connection between *Measurement Independence* and the idea of free will, thus casting doubts on *Measurement Independence* as an adequate requirement in the derivation of the Bell inequalities, it may be pointed out that the discussion—and therefore the conclusions as well– misses the point. For, it may be argued, it is not *free will* actually what *Measurement Independence* stands for but rather a more general idea related to the lack of a world (or cosmic) *conspiracy* (by which EPR measurement settings would be pre-established by a hidden variable in their past history).[4] More precisely, one can claim that it is not free will, in fact, what is behind the justification of Equation (8.1), but just the intuition that the world is not such that it conspires to pre-set measurement choices in an EPR experiment, regardless of how are those actually decided, i.e. be it by means of random radioactive devices, lottery boxes, some computer routine involving random numbers with no need for human interaction or operation, or even by means of human (free) decisions. In this view, thus, lack of free will would only be one very specific way among other possibilities a world conspiracy may show-up in a theory.

Under this more general approach we need to concentrate then on the alleged correspondence between *Measurement Independence* and the lack of a world conspiracy. Just like before, we may ask, does *Measurement Independence* need to be required in order to exclude a world conspiracy?

Again, and just like in the case of free will as a justification for *Measurement Independence*, the question above is very seldom addressed in the literature. I shall tackle the issue only briefly here. In particular, I shall generalise one of the arguments offered in San Pedro (2013), as noted above, in relation to *Measurement Independence* and free will to claim that requiring the world to be free of conspiracies does not necessarily mean that the condition of *Measurement Independence* is to be in place. The actual argument, as we shall see below, exploits

---

[4]This is, by the way, where the origin of the terminology *No-conspiracy* can be traced back to.

two specific background assumptions behind the requirement of *Measurement Independence*, related to the space-time arrangement of events in the EPR scenario. As a consequence, I shall conclude that it is not the case that violations of *Measurement Independence* entail any sort of world conspiracy, and therefore by contraposition that we can make sense of (common cause) causal pictures of EPR which feature such violations explicitly.

8.3   Temporal Order and Causal Order

Measurement Independence, as we have seen, is a probabilistic independence requirement with a straightforward causal grounding (and interpretation). Namely that common causes C postulated to explain the EPR correlations be causally independent of measurement operations m_i which lead to the correlated outcomes O_i. This interpretation however involves further assumptions related among other things to the specific time arrangement of events in an EPR experiment as well as, more generally, to how causal relations are to be understood.

(This last sentence is deliberately left vague. There are a number of causal assumptions onto which the referred interpretation of *Measurement Independence* is grounded. For instance, one needs to assume that causal relations, or the lack of them, are faithfully represented by probabilistic expressions, be it correlations in case of causal dependence, or probabilistic independence in case of causal independence. In what follows however I shall only concentrate on one such assumption, specifically related to the direction of causation. The reason for this is that, as we shall see, it is this particular assumption that can help us make sense of the other presupposition discussed here, about the temporal arrangement of events in the EPR scenario.)

Starting with the more specific assumptions behind *Measurement Independence*. The causal independence between the postulated common cause C and the measurement operations m_i presuppose a particular (fixed) time ordering of the events involved. More precisely, common causes are assumed to take place *before* measurement operations do (and therefore before any outcome is registered).

Let us call this presupposition *Time Order*:

Presupposition (Time Order) *The temporal arrangement of events in EPR is such that postulated common causes C take place* before *measurement operations m_i in both wings of the experiment do.*

The meaning of 'before' above amounts to an event 'being in the past' of the other, i.e. laying within the corresponding backwards light-cone. The actual time arrangement presupposed by *Time Order* results in the light-cone structure of Fig. 8.1.

The presupposition of *Time Order* above is rooted in the intuition that common causes are just hidden variables aimed at completing the otherwise incomplete description of the EPR phenomena offered by quantum mechanics—and by

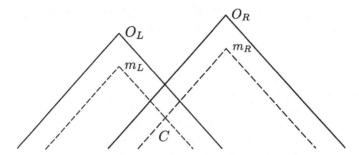

Fig. 8.1 *Time Order* demands that common causes C be in the past of measurement operations m_i ($i = L, R$) in both wings of the EPR set-up (and therefore in the past of measurement outcomes O_i ($i = L, R$) as well)

extension, therefore of any quantum phenomena.[5] As such, they are supposed to be a "missing" part of the quantum mechanical description of the singlet state, i.e. some missing bit of the actual (real) singlet state itself. This so close a relation to the actual singlet state is what seems to warrant the assumption that such hidden variables, i.e. common causes, need to be spatio-temporally located, if not at the very same source—where the singlet state sits—, in its (very) close vicinity. That is, in the intersection of the measurement operations' backwards light-cones.

As it happens *Time Order* is rarely discussed in the literature and very often assumed only implicitly. I shall suggest in a moment however that the particular temporal arrangement of events presupposed is not the only available option. In particular, causal models can be conceived in which this particular temporal order of events is altered.

Before going through such examples, let us discuss a further more general assumption behind *Measurement Independence*. It has to do with how actual causal relations propagate, and more precisely with the direction of causation. Indeed, for the above interpretation of *Measurement Independence* to make sense at all one needs to presuppose that causes *always* lie in the past of their effects. Let us call this presupposition *Causal Order*:

Presupposition (Causal Order) *Causes propagate* always *forward in time, i.e. causes* always *precede (temporally) their effects.*

Causal Order as formulated above seems closely related to our previous *Time Order* presupposition. In fact, they are related but only in the sense that presupposing *Time Order* in the usual attempts to provide causal explanations of EPR correlations does only make sense in the context of *Causal Order*. In other words, the particular temporal arrangement of events demanded in *Time Order* is a consequence of the assumption that *all* causes propagate forward in time. This is

---

[5]This is indeed the view defended by Bas van Fraassen in his influential "Charybdis of Realism …" (van Fraassen 1982).

not however a *logical consequence*. Indeed, the two presuppositions are logically independent and, as we will see, can each hold or fail regardless of the other. Failure of any of them, or both, will also entail a failure of *Measurement Independence* in any case.

What *Causal Order* practically does is to rule out *any* possibility of backwards in time causation in our causal picture, whatever the particular time arrangement of events is. It is in fact what our most robust commonsense intuitions about causal relations seem to recommend. In the particular case of the EPR correlations then, *Causal Order* bans the postulated common causes to influence events in their past, or else be influenced by any events in their future—such as for instance measurement operations (in case *Time Order* is also in place, of course). Once more, this is usually taken to be the correct and most natural way to think about causal influences in the EPR scenario. Some authors disagree however and prefer to leave open the possibility that some causal relations may propagate backwards in time.[6]

In sum, requiring *Measurement Independence* involves a combination of an assumption about the temporal order of events, i.e. *Time Order*, as well as an assumption about the direction of causal influences, i.e. *Causal Order*.

8.4   Three Common Cause Models

As noted above, *Time Order* and *Causal Order* are logically independent statements which can each hold or fail on their own grounds, regardless of the other. And in fact rejecting either of them separately, or both of them at the same time, yields at least three different causal pictures or models, as suggested in San Pedro (2013). Obviously, all such causal models will violate *Measurement Independence*. Table 8.1 (page 154) displays the logical structure, when it comes to the *Time Order*

Table 8.1 Common cause models where *Measurement Independence* is not satisfied do violate either *Time Order* or *Causal Order* presuppositions, or both. The standard common cause structure assumed in the usual derivations of Bell's theorem (i.e. 'Standard Common Cause Model'), in contrast, satisfies both of these, and of course *Measurement Independence* as well

Time and causal order in common cause models			
	Measurem. Ind.	*Time order*	*Causal order*
CC Model 1	x	✓	x
CC Model 2	x	x	✓
CC Model 3	x	x	x
Stand. CC Model	✓	✓	✓

---

[6]As we shall see in a moment, a violation of *Causal Order* will not mean that *all* causal influences propagate backwards in time but just that *some* do. This will be in fact the source of some criticism as regards common cause models featuring a violation of *Causal Order*.

and *Causal Order* presuppositions, of each of the three common cause models discussed below, plus that corresponding to the common cause model assumed in the usual approaches, which I refer to as 'Standard Common Cause Model'.

Common Cause Model 1: A first causal picture results from keeping the temporal arrangement of events assumed in standard treatments of causal explanations of the EPR correlations, i.e. what I called *Time Order* above, but rejecting the assumption that causal influences propagate *only* forward in time, i.e. *Causal Order*.

In other words, one may assume that, as usual, postulated common causes take place before *both* measurement operation events (in both wings), as well as of course before the outcome events. Again, that this is the right temporal sequence of events which a given run of an EPR experiment must feature is normally presupposed without further justification. Yet I cannot see any particular reason why this needs to be so.[7] With the *Time Order* assumption in place, nevertheless, the intuition is that violations of *Measurement Independence* can only make sense—with no world conspiracies involved—if the corresponding causal model features backwards in time causation.

In this picture then actual measurement operations m_i are (future) causes of the postulated common causes C, which are in turn causes of the measured outcomes O_i. Provided the common cause C is postulated to be an event in the remote past of both outcomes, i.e. an event located somewhere in the overlap of the outcomes backwards light-cones, all causal influences in the model will be completely local influences, hence avoiding conflict with special relativity (see Fig. 8.2).

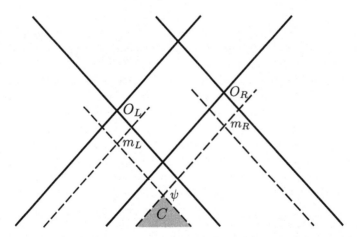

Fig. 8.2 Backward light-cone structure for measurement operations propagating causally backwards in time to influence the postulated common causes in their past. Common causes, in turn propagate, as usual forward in time to cause corresponding future outcomes

---

[7]However, rejecting *Time Order* would result in other different causal pictures (see causal models 2 and 3 below).

Problems with such a model will come obviously from the fact that some of the causal influences they feature propagate backwards in time. On the one hand, backwards causation is not really a preferred option in the relevant literature, as it is taken by most to be highly counter-intuitive. It is fair to say that some quite influential authors still keep this option open.[8]

This kind of causal models will also face difficulties related to the fact that *only some* of the causal influences in them propagate backwards in time. In particular, for such models to work they need to display a combination of causal influences propagating in both directions (of time), i.e. both backward and forward causal influences. (Note that saying that the model is a backwards causation model does not mean that *all* causal influences in it take place backward in time. To ensure that all causal influences took place backwards in time we would need to change— reverse—the time order of all events, which would probably take us to a full backwards in time causal model in which *Measurement Independence* would not be violated.)

The question is then, how can one tell, in a particular causal structure, which causal relations should actually take place forward time and which backward in time (and why these and not some others). Is seems, in particular, there is no way to identify in advance what events will feature causal influences in one or the other direction, i.e. there is no way to know when to expect one or the other. In the EPR case above, for instance, why should measurement operation events m_i, and not some other event, have the "privilege" of being able to propagate their causal influences backward in time?

Further worries may arise as well in relation to the possibility of facing paradoxical situations if we admit both forward and backward causal influences.[9] The causal picture above, for instance, leaves unspecified what kind of causal influences may the experiment's outcomes be associated to, i.e. whether they are causes propagating forward or backward in time. It would be perfectly fine not to worry about this if the outcomes would operate causally as usual, i.e. forward in time. However once we admit the possibility that one of them may propagate causally backwards in time we are open to paradox. In particular, it is perfectly conceivable in that case, say, that O_L in Fig. 8.2 has some causal influence on m_L, i.e. that an outcome has some causal influence on the measurement setting that gave rise to it in the first place.[10]

This would have undesired consequences. On the one hand, it would be a clear threat to the attempt of the model to save free will since the experimenter's choices when setting the measurement apparatus would no longer be free—they would be

---

[8]Classical references to retrocausal pictures include Sutherland (1983), de Beauregard (1987) or more recently Price (1994, 1996). Huw Price, for instance, has gone as far as to argue that the characteristic time-symmetry of quantum mechanics (and of microphysics in particular) may imply, given some further assumptions about the ontology of the theory, the existence of backwards in time causation, or retrocausality, as he terms it (Price 2012).

[9]I must thank a reviewer for hinting this further kind of difficulties.

[10]Note that we cannot rule out this possibility completely just by looking at the other relevant probabilistic relations among O_L and m_L. In fact, we should expect them both to be correlated.

caused by the outcome. On the other hand, allowing O_L to causally influence m_L would result in the model featuring a causal loop between these and the common cause C. All these are problems that any retrocausal model of EPR correlations needs to address.

Worries as the above also reveal the close relation between *Time Order* and *Causal Order* and suggest further causal models resulting from a violation of the requirement that the time arrangement of events follows the specific structure demanded in *Time Order*. So, why not considering actual violations of *Measurement Independence* via a failure of (at least) *Time Order*? In other words, why not revise the particular time arrangement of events fixed by *Time Order*, and consider possible modifications of it? This would lead to at least two further causal models.

Common Cause Model 2: As a first option, we could consider causal structures which violate *Measurement Independence* via a failure of *Time Order* only. That is, causal models built without the premise that the postulated common cause C be in the past of both measurement operations m_i and corresponding outcomes O_i, while keeping the intuition that, in effect, causal influences propagate exclusively forward in time, i.e. satisfying *Causal Order* above.

A causal model along these lines is discussed in San Pedro (2012), for instance. There, a common cause C is postulated to take place just in between the actual measurements m_i and the corresponding outcomes O_i. That is, the common cause is now taken to lie in the future of measurement operations m_i but in the past of the resulting EPR outcomes O_i.[11] In terms of space-time structure, the model postulates a common cause which ought to be located somewhere in the union of the two regions defined by the double light-cones formed by each measurement operation m_i and the corresponding outcome O_i (see Fig. 8.3).[12]

With this new temporal arrangement of events therefore, measurement operations can well be thought to be causally relevant for the postulated common cause, which in turn would be responsible of such and such outcome being observed. And all causal influences propagate forward in time. This has as a consequence that some of the causal influences in the model will clearly turn out to be non-local. This becomes obvious from the fact that, as already pointed out, postulated common causes C shall now be located somewhere in the union of the double light-cones formed by the measurement operations m_i and the corresponding outcomes O_i (see again Fig. 8.3). And since these two regions are space-like separated any causal influence from any of them on the distant wing outcomes will forcefully be non-local.

---

[11]Obviously, this is not the only temporal arrangement possible. *Time Order* could be violated as well by altering the order between measurements and outcomes, and leaving the common cause in the past of both, just as it is standard. The problem with such a structure, of course, is that it is very difficult to think of outcomes taking palace (or being observed) before measurements have been performed.

[12]This model, of course, needs to assume that there is some lapse of time—or rather some region of space-time—, however little this may be, between the performance of a measurement and the occurrence of the corresponding outcome. See San Pedro (2012) for details.

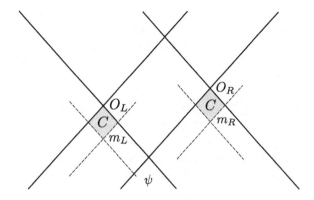

Fig. 8.3 Backward light-cone structure for common causes C located in the future of measurement events m_i, but in the past of resulting outcomes O_i. All causal influences propagate, as usual, forward in time

Quantum non-locality is not new however, and one may even argue that it is not really such a strange feature any more. Critics may want to note nevertheless that simpler non-local explanations of the EPR correlations are already available without the need to refer to the notion of common cause at all—such as non-local direct causal models, for instance. In other words, why do we need common causes for the explanation of EPR correlations at all, if they just seem to constitute a further unnecessary complication?

It is worth noting, in response to such criticism, that common cause models as the above might have more to offer than simpler direct cause models. In the case of the model outlined here, for instance, the fact that measurement operations are taken to be explicit causal relevant factors for the EPR outcomes can be seen as the model telling us something about measurement in quantum mechanics. Namely, that quantum measurement is a causal, generally non-local, process.

Finally, let me point out that endorsing a model along these lines can also be motivated by some results in algebraic quantum field theory (AQFT). In particular, for instance, it has been recently shown that (the equivalent to) common cause events postulated to explain distant correlations in AQFT exist, but these must be located in the union of the correlated events backward light-cones, i.e. a so-called "Weak Reichenbach's Common Cause Principle" holds in AQFT (Rédei and Summers 2002). This of course does not mean that the common cause events be necessarily located precisely in the shaded double-cone regions in Fig. 8.3 above. However, that this may be so certainly remains an open possibility.

Common Cause Model 3: Rejecting both *Time Order* and *Causal Order* at the same time results in yet another causal structure displaying backwards in time causation (just as in the first causal model discussed above). In this case however the presence of backwards in time causal influences will only make sense if the alteration of the temporal arrangement of events originally fixed by *Time Order* is just different than the one assumed in our previous Common Cause Model 2. In particular, putative common causes C need now be postulated as events located in the future of both measurement operations m_i and corresponding outcomes O_i (see Fig. 8.4).

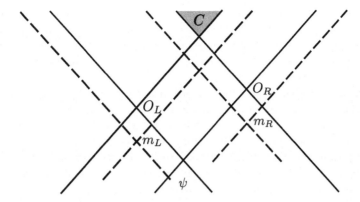

Fig. 8.4 Backward light-cone structure for common causes in the future of both measurement operations and outcome events. Common causes operate backwards in time to cause corresponding events

Also as in the first model above, causal influences shall now turn out to be completely local as long as the postulated common cause is located far enough in the future of the experiment outcome events. Locality will be assured, more in particular, if the common cause C is assumed to lie in the overlap of the EPR outcomes future light-cones (see again Fig. 8.4). We can see then, once more, that locality is achieved at the cost of introducing backwards in time causation in our models. Just as in Model 1, this causal picture can avoid conflict with special relativity fairly easily.

Obviously, we shall now also face similar problems to the ones encountered with Model 1, related precisely to backwards causation. Namely, we shall face the charge that backwards in time causal influences are highly counterintuitive. But also, and just as with Model 1, we shall now face the fact that cannot provide a satisfactory account of how backward in time influences can be identified precisely. In particular in this model the causal structure also features now a combination of forward and backward causal influences. And again there does not seem a good way to tell what this precise combination is, i.e. which events exactly will propagate causally forward in time, and which will do it backwards. Once more the only reply available to such criticism seems to be that it is the specific temporal arrangement of events that the model presupposes what fixes the specific combination of events propagating either forward or backward in time.

8.5 Conclusion

I have revised above three common cause models all of which violate *Measurement Independence*, and therefore avoid the implications of Bell's theorem as regards the existence of common cause explanations of EPR correlations. The three models were obtained each as a result of rejecting two further presuppositions behind *Measurement Independence*. Namely, that the temporal arrangement of events is a given specific one—that in Fig. 8.1—, which needs to be kept fixed, i.e. *Time Order*, and that the direction of causal influences is also fixed (and taken to be forward in time), i.e. *Causal Order*.

The discussion above highlighted, in each case, a particular aspect of the quantum description of the EPR experiment and/or its phenomenology. For instance, causal pictures in which *Causal Order* was violated, i.e. Models 1 and 3, showed that the tension between causality and non-locality (most commonly taken to be expressed by the implications of Bell's theorem) can be dissolved once we admit the possibility that backwards in time causal influences may take place, and regardless of the specific space-time arrangement of the events involved. However, backwards in time causation suffers from its own problems.

On the other hand, preserving our usual intuitions about the direction of causation, but allowing instead that the temporal arrangement of events fixed by *Time Order* be altered results in a causal structure where some of the causal relations happen to be non-local. Despite the well known difficulties that non-locality introduces, such a model might be useful to investigate the precise (causal) role of measurement in quantum mechanics.

To close, aside the above specific issues the three models discussed here raise, what they show, above all, is that it makes perfect sense to consider violations of *Measurement Independence* which do not convey in any way world conspiracies, or conflict with us having freedom of will at all.

References

Bell, J. 1964. On the Einstein-Podolsky-Rosen paradox. *Physics* 1: 195–200. Reprinted in Bell 1987, pp. 14–21.

Bell, J. 1987. *Speakable and unspeakable in quantum mechanics*. Cambridge: Cambridge University Press.

Cushing, J., and E. McMullin, eds. 1989. *Philosophical consequences of quantum theory*. Notre Dame: University of Notre Dame Press.

de Beauregard, O.C. 1987. On the zigzagging causality EPR model: Answer to Vigier and coworkers and to Sutherland. *Foundations of Physics* 17: 775–785.

Graßhoff, G., S. Portmann, and A. Wüthrich. 2005. Minimal assumption derivation of a Bell-type inequality. *The British Journal for the Philosophy of Science* 56: 663–680.

Hofer-Szabó, G., M. Rédei, and L.E. Szabó. 2002. Common causes are not common-common causes. *Philosophy of Science* 69: 623–636.

Price, H. 1994. A neglected route to realism about quantum mechanics. *Mind* 103: 303–336.

Price, H. 1996. *Time's arrow and Archimedes' point*. New York/Oxford: Oxford University Press.

Price, H. 2012. Does time-symmetry imply retrocausality? How the quantum world says "maybe"? *Studies in History and Philosophy of Modern Physics* 43: 75–83.

Rédei, M., and S. Summers. 2002. Local primitive causality and the common cause principle in quantum field theory. *Foundations of Physics* 32: 335–355.

Reichenbach, H. 1956. *The direction of time*, ed. M. Reichenbach. Unabridged Dover, 1999 (republication of the original University of California Press 1956 publication).

San Pedro, I. 2012. Causation, measurement relevance and no-conspiracy in EPR. *European Journal for Philosophy of Science* 2: 137–156.

San Pedro, I. 2013. On free will and no-conspiracy. In *New vistas on old problems: Recent approaches to the foundations of quantum mechanics*, eds. T. Sauer and A. Wüthrich, 87–102. Berlin: Edition Open Access, Max Planck Research Library for the History and Development of Knowledge.

Sutherland, R.I. 1983. Bells theorem and backwards-in-time causality. *International Journal Theoretical Physics* 22: 377–384.

van Fraassen, B.C. 1982. The charybdis of realism: Epistemological implications of Bell's inequality. *Synthese* 52: 25–38. Reprinted with corrections in (Cushing and Mcmullin, 1989, pp. 97–113).

Part III
Indeterminism, Undecidability, and Macrostates

Chapter 9
Meaning, Truth, and Physics

László E. Szabó

Abstract A physical theory is a partially interpreted axiomatic formal system (L, S), where L is a formal language with some logical, mathematical, and physical axioms, and with some derivation rules, and the semantics S is a relationship between the formulas of L and some states of affairs in the physical world. In our ordinary discourse, the formal system L is regarded as an abstract object or structure, the semantics S as something which involves the mental/conceptual realm. This view is of course incompatible with physicalism. How can physical theory be accommodated in a purely physical ontology? The aim of this paper is to outline an account for meaning and truth of physical theory, within the philosophical framework spanned by three doctrines: physicalism, empiricism, and the formalist philosophy of mathematics.

9.1 Introduction

1. All the forthcoming considerations will be based on the following three philosophical premises.

Physicalism: everything is physical; all facts supervene on, or are necessitated by, the physical facts.

Empiricism: genuine information about the world can be acquired only by a posteriori means.

Formalism: logic and mathematics are thought of as statements about manipulations with meaningless symbols.

I won't argue for these doctrines in this paper – they are legitimate philosophical positions. I take them as initial premises. Rather, I will discuss a few radical consequences of them, concerning the fundamental nature of logic and mathematics, and the structure of physical theories. Nevertheless, it must be mentioned that the ontological doctrine of physicalism is of primary importance, which, in some

L.E. Szabó (✉)
Department of Logic, Institute of Philosophy, Eötvös Loránd University Budapest,
Budapest, Hungary
e-mail: laszlo.e.szabo@gmail.com

© Springer International Publishing AG 2017
G. Hofer-Szabó, L. Wroński (eds.), *Making it Formally Explicit*, European Studies
in Philosophy of Science 6, DOI 10.1007/978-3-319-55486-0_9

sense, implies both formalism and empiricism. The first implication is rather trivial. Physicalism denies the existence of mental and abstract entities; consequently, there is no room left for any kind of platonism or mentalism in the philosophy of logic and mathematics. Therefore, disregarding a Millian-style physical realist approach in which mathematics itself becomes a physical theory, in the sense of physical theory as described later in this paper, formalism – more precisely, what we will call physico-formalism – seems to be the only account for logic and mathematics that can be compatible with physicalism. Regarding the second implication, we will see in point **10** that empiricism will be a consequence of our physico-formalist account of the semantics of physical theory.

In the first section I will outline the basic ideas of physico-formalist philosophy of mathematics. In the second section, combining the physico-formalist approach to formal system with an intuition we can learn from Gödel's proof of incompleteness theorem, I will sketch a physicalist–formalist–empiricist – or shortly, physicalist – theory of meaning and truth with respect to physical theories.

9.2 Physico-Formalist Philosophy of Mathematics

2. Physico-formalist interpretation of mathematics (Szabó 2003, 2012) is a reflection to the following fundamental problem: If physicalism is true, then the logical/mathematical facts must be necessitated by the physical facts of the world. This raises the question: How can the logical and mathematical facts be accommodated in a purely physical ontology? In order to answer this question, first we need to clarify: What is it that has to be accounted for within a physicalist ontology? What are the logical/mathematical facts?

Our starting point is the formalist thesis. In Hilbert's famous formulation it says:

> **The formalist thesis:** Mathematics is a game played according to certain simple rules with meaningless marks. (Bell 1951, 38)

In other words, a mathematical statement/fact/truth is like "$\Sigma \vdash A$" – with *single* turnstile, it must be emphasized. Note that this claim is more radical than, as sometimes called, "if-thenism", according to which the truth of A is based on the assumption that the premises in Σ are true. According to the formalist view, neither A nor the elements of Σ are statements, which could be true or false. They are just strings, formulas of the formal system in question. Derivation is not a truth preserving "if-then"-type reasoning as traditionally taken, but a mechanistic operation with meaningless strings of symbols.

3. This is the point where *physico*-formalist philosophy of mathematics starts. We turn to the following question: Where are the states of affairs located in the ontological picture of the world that make propositions like "$\Sigma \vdash A$" true or false? The physico-formalist answer is this:

Fig. 9.1 A formal system represented in a laptop with an inserted CD. The CD contains a program making the computer to list the theorems of the formal system

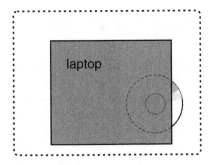

> **The physico-formalist thesis:** A formal system should be regarded as a physical system which consists of signs and derivational mechanisms embodied in concrete physical objects, concrete physical configurations, and concrete physical processes.

Therefore, a "$\Sigma \vdash A$"-type mathematical proposition expresses an *objective fact* of the formal system as a particular portion *of the physical world*.

One can argue for this physicalist account of formal system in three steps. (For more detailed argumentation see Szabó 2012).

(I)   A formal systems can be represented in a physical system. Imagine a laptop with an inserted CD (Fig. 9.1). The CD contains a program completely determining the laptop's behavior: in some order, the computer lists the theorems (together with the proofs) of a formal system. I think, it is commonly accepted to say that in the "computer + CD" system we have "a physical representation of the formal system" in question. In this representation, "$\Sigma \vdash A$" (that is, formula A is printed) becomes a fact of the physical reality inside the dotted line.

(II)   All mathematical facts can be thought of as a physical fact in some physical representation. It is because, in Curry's words, "in order to think of a formal system at all we must think of it as represented somehow". (Curry 1951, 30) This is in agreement with the widely accepted view in the recent literature on physical computation that "it is only physical objects, such as computers or human brains, that ever give us glimpses of the abstract world of mathematics". (Deutsch et al. 2000, 265)

(III)   Actually, there is nothing to be "represented"; there is nothing beyond the flesh and blood formal systems. Consider the context of the above quoted sentence of Curry:

> [A]lthough a formal system may be represented in various ways, yet the theorems derived according to the specifications of the primitive frame remain true without regard to changes in representation. There is, therefore, a sense in which the primitive frame defines a formal system as a unique object of thought. This does not mean that there is a hypostatized entity called a formal system which exists independently of any representation. On the contrary,

in order to think of a formal system at all we must think of it as represented somehow. But when we think of it *as* formal system we abstract from all properties peculiar to the representation. (Curry 1951, 30)

But, what does such an "abstraction" actually mean? What do we obtain if we abstract from some unimportant, peculiar properties of a *physical* system L constituting a "physical representation of a formal system"? In order to think of this abstraction at all, in order to differentiate the important and unimportant features of the physical system L, and to change from a more detailed description of the system to a less detailed one, we must have a physical theory (M, S) – in the sense of the definition of physical theory in the next section – describing the physical system L in question. However, the formal system M also is "represented somehow", in Curry's words; it is another flesh and blood formal system. That is to say, the whole abstraction – the result of the abstraction included – is contained in another flesh and blood formal system M. So, instead of obtaining a non-physical "abstract formal system" we remain to have flesh and blood formal systems.

Another way to think of this abstraction is to consider what is common to the different "physical representations", that is, to describe the common features of different flesh and blood formal systems $L_1, L_2, \ldots L_n$. But, to describe the common features of physical systems $L_1, L_2, \ldots L_n$ we have to have a physical theory (M, S) which is capable of describing all $L_1, L_2, \ldots L_n$ together (Fig. 9.2). Only in a suitable formal system M it is meaningful to talk about similarity or isomorphism between the structures describing $L_1, L_2, \ldots L_n$, and about the *equivalence class* of these structures, which could be regarded as an "abstract formal system". But, all these

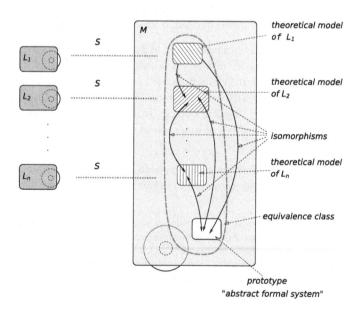

Fig. 9.2 To describe the common features of physical systems $L_1, L_2, \ldots L_n$ we have to have a physical theory (M, S) which is capable of describing all $L_1, L_2, \ldots L_n$ together

objects live in the formal system M which also is "represented somehow", that is, in a formal system existing in the physical world.

Thus, abstraction, properly understood, does not lead out of the physical realm. It does not produce "abstract formal systems" over and above the physically existing "representations". Actually, there is nothing to be "represented"; there is nothing beyond the flesh and blood formal systems. That is to say, a formal system *"as formal system"* is a part of the physical reality. Consequently, any statement about a formal system – including a statement like "$\Sigma \vdash A$" – is a statement of a physical fact; and it has exactly the same epistemological status as any other statements about the physical world.[1]

4. There are far-reaching consequences of the above considerations concerning an old problem of empiricism. In Ayer's words:

> Where the empiricist does encounter difficulty is in connection with the truths of formal logic and mathematics. For whereas a scientific generalization is readily admitted to be fallible, the truths of mathematics and logic appear to everyone to be necessary and certain. But if empiricism is correct no proposition which has a factual content can be necessary or certain. Accordingly the empiricist must deal with the truths of logic and mathematics in one of the following ways: he must say either that they are not necessary truths, in which case he must account for the universal conviction that they are; or he must say that they have no factual content, and then he must explain how a proposition which is empty of all factual content can be true and useful and surprising. ...
>
> If neither of these courses proves satisfactory, we shall be obliged to give way to rationalism. We shall be obliged to admit that there are some truths about the world which we can know independently of experience; ... (Ayer 1952, 72.)

The physico-formalist approach resolves this difficulty in the following way:

1. Logical and mathematical truths express objective (mind independent) facts of a particular part of the physical world, namely, the facts of the formal systems themselves. As such, they are synthetic, a posteriori, not necessary, and not certain; they are fallible.
2. But they have contingent factual content, as any similar scientific assertion, so they "can be true and useful and surprising". The logical and mathematical facts can be discovered, like any other facts of nature, just like a fact about a plastic molecule, or other artifact.
3. The fact that the flesh and blood formal systems usually are simple physical systems of relatively stable behavior, like a clockwork, and that the knowledge

---

[1]Let me emphasize the distinction between the physico-formalist interpretation of mathematics and the so called immanent or physical realism in the sense of Mill. For example, '$a^2 + b^2 = c^2$' is a theorem of Euclidean geometry, $\{$Euclidean$\} \vdash a^2 + b^2 = c^2$. According to immanent realism, '$a^2 + b^2 = c^2$' reflects some general truth about the real triangles in the physical world. Or at least, in the sense of the structuralist version of immanent realism, the axiomatic theory "Euclidean geometry", as a whole, expresses some true structural property of the physical world; something about the congruence of rigid bodies, or the likes. The physico-formalist theory also claims that $\{$Euclidean$\} \vdash a^2 + b^2 = c^2$ expresses something in the physical world. But this something has nothing to do with the physical triangles and rigid bodies. $\{$Euclidean$\} \vdash a^2 + b^2 = c^2$ expresses a property of the formal system "Euclidean geometry", a property of the physical system consisting of the signs and the derivation mechanisms.

of logical and mathematical truths does not require observations of the physical
world external to the formal systems explains the universal *illusion* that logical
and mathematical truths are necessary, certain and a priori.

Thus, empiricism is not challenged by the alleged necessary truths delivered by
logical and mathematical reasoning. On the contrary, consequent physicalism can
resolve the long-standing debate surrounding the truth-of-reasoning versus truth-
of-facts dichotomy. Logical and mathematical truths are nothing but knowledge
obtained from experience with respect to particular physical systems, the formal
systems themselves. Since logical and mathematical derivations are "reasoning"
par excellence, one must conclude that reasoning does not deliver to us necessary
truths. Logical and mathematical reasoning is, if you like, an experiment with a flesh
and blood formal system. Therefore, we must draw the epistemological conclusion:
There is no *higher* degree of certainty than available from experience.

5. Formal systems are man-maid physical objects. As such, they are not eternal
and not readily given to us. Nevertheless, it is of course not true that anything
can be created with arbitrary properties. The objective features of physical reality
predetermine what can be created and what cannot. For example, even if we assume
that there are no polyvinyl chloride molecules in the universe, except the ones
created by mankind, the laws of nature admit the possibility of their existence and
predetermine their properties. Similarly, the laws of nature predetermine what kinds
of formal system can exist. In this sense, the logical and mathematical facts are
eternal and independent from us. They are contingent as much as the laws of nature
are contingent. In other words, Paracelsus' God can make "a triangle with four
sides" exactly as well as "an ass with three tails".[2]

9.3   Physical Theory

6. Following Carnap, a physical theory can be considered as a partially interpreted
formal system, (L, S), providing a description of a certain part of physical reality,
U. In general, the system of axioms of L contains some logical axioms and

---

[2]Notice that this claim of contingency of logical and mathematical facts is quite different from what
we call mathematical contingentism in the debates about contingency vs. necessity of the existence
of mathematical entities (e.g. Field 1993; Colyvan 2000; Miller 2012). In the physico-formalist
approach a formula of a formal system carries no meaning; Platonic, fictional, whatsoever. The
meaningful statements of logic and mathematics, which can be true or false, are of form "$\Sigma \vdash A$".
They do refer, indeed, to real facts of a real object: the physically existing formal system. This
object, as we will see in the next section, is indeed indispensable for our physical theories. (Cf.
Colyvan 2004)

the derivation rules (ideally, the first-order predicate calculus with identity), the axioms of some mathematical theories, and, of course, some physical axioms. (The distinction between "logical", "mathematical", and "physical" are rather terminological/traditional than essential.) The semantics S is understood as a kind of correspondence between (some of) the formulas of L and the states of affairs in U. Consider an interpreted formula A in L. One has to distinguish between the following two notions:

1. *A is a theorem* in L, $L \vdash A$.
2. *A is true*, that is, according to the semantics S, A refers to a state of affairs in U, which is in fact the case.

For example, 'The electric field strength of a point charge is $\frac{kQ}{r^2}$' is a theorem of Maxwell's electrodynamics – one can derive it from the Maxwell equations. On the other hand, according to the semantics relating the symbols of the Maxwell theory to the empirical terms, this sentence corresponds to an empirically observable fact of the physical reality.

A physical theory is considered to be true if these two things are in harmony: if A is a prediction of the theory, $L \vdash A$, then it corresponds to a state of affairs which is indeed the case in U. By and large, this is the standard picture of a physical theory.

7. Let us turn to the ontological issues. It is perhaps not far from the truth to say that the standard view regards the formal system L as an abstract object or structure, the semantics S as something which involves the mental/conceptual realm, and, of course, U as a part of the physical world. This view is of course incompatible with physicalism. How can physical theory be accommodated in a purely physical ontology? The L-part is already solved by the physico-formalist interpretation of formal system and logical/mathematical fact. But, how can the physicalist account for *meaning* and *truth*? Again, first we need to clarify what it is that has to be accounted for; we need a definition of semantic relationship between formulas of a formal system and states of affairs in the physical world.

The definition will be based on the intuition we can learn from Gödel's construction of representation of the meta-arithmetic facts in Peano arithmetic, in the preparation of the first incompleteness theorem (e.g. Crossley et al. 1990, 62). Let $Pr(x, y)$ denote the meta-arithmetic fact that the formula–sequence of Gödel number x constitutes a proof of the formula of Gödel number y. According to Gödel's construction, every such meta-arithmetic fact $Pr(x, y)$ is represented with a PA formula $R(x, y)$ if the following condition is met: For all Gödel numbers (x, y),

$$\text{if } Pr(x, y) \text{ is the case then } \Sigma_{PA} \vdash R(x, y)$$

$$\text{if } Pr(x, y) \text{ is not the case then } \Sigma_{PA} \vdash \neg R(x, y)$$

where Σ_{PA} denotes the axioms of Peano arithmetic.

Mutatis mutandis: one is entitled to say that a formula A represents or means a state of affairs a in U, if the following two conditions are met:

(a) There exist a family $\{A_\lambda\}_\lambda$ of formulas in L and a family $\{a_\lambda\}_\lambda$ of states of affairs in U, such that $A = A_{\lambda_0}$ and $a = a_{\lambda_0}$ for some λ_0.
(b) For all λ,

$$\text{if } a_\lambda \text{ is the case in } U \text{ then } \Sigma \vdash A_\lambda$$

$$\text{if } a_\lambda \text{ is not the case in } U \text{ then } \Sigma \vdash \neg A_\lambda$$

8. Turning to the ontological issues, it must be clear that a_λ – as a symbol in the meta-language we use to describe the semantic relationship – stands for a state of affairs, a configuration of the physical world. It is not a linguistic object, it is not a "sentence of the theory expressing the state of affairs" in question. It is A_λ that is a linguistic object, a formula of L, that can refer to a_λ, given that condition (b) holds.

According to the physico-formalist approach "$\Sigma \vdash A_\lambda$" and "$\Sigma \vdash \neg A_\lambda$", respectively, are states of affairs in the physical world, facts of the flesh and blood formal system L. Thus, what we observe in condition (b) is a kind of regularity or correlation between physical facts of two parts of the physical world, L and U. More precisely, it should be a real correlation. To explain what "real" means here, let me give a simple example. Imagine a computer programmed such that entering a random number $z \in (0, 90)$ it calculates by applying some laws of physics whether the bullet shot from the gun will land in the plate or not if the angle of inclination is z (Fig. 9.3). At the same time we perform the experiment with a randomly chosen angle of inclination $\alpha \in (0, 90)$. Now, imagine a laboratory record as in Table 9.1. In this table, the pairing between the facts of the computer and the facts of the experiment was based on simultaneity. It is based on a real physical circumstance establishing a real conjunctive relation between the elements of the two families. There is no correlation; condition (b) is not satisfied.

One can however *conceive* another parametrization of the two families as is shown in Table 9.2, according to which condition (b) is satisfied. Are we entitled to say that the computer makes real correct predictions? That a given "Yes" is a prediction of "Buuum!"? Of course not, because this parametrization is only a conceived one; it does not establish a real conjunctive relationship.

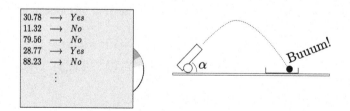

Fig. 9.3 Imagine a computer programmed such that entering a random number $z \in (0, 90)$ it calculates whether the bullet will land in the plate or not if the angle of inclination is z. At the same time we perform the experiment with a randomly chosen angle of inclination $\alpha \in (0, 90)$.

Table 9.1 Pairing by simultaneity

α	Outcome	z	Result
88.23	–	30.78	Yes
79.56	–	11.32	No
30.78	Buuum!	79.56	No
11.32	–	28.77	Yes
28.77	Buuum!	88.23	No
29.02	Buuum!	62.25	No
62.25	–	34.21	Yes
36.54	Buuum!	29.02	Yes
34.21	Buuum!	36.54	Yes

Table 9.2 A conceived pairing

α	Outcome	z	Result
88.23	–	11.32	No
79.56	–	79.56	No
30.78	Buuum!	30.78	Yes
11.32	–	88.23	No
28.77	Buuum!	28.77	Yes
29.02	Buuum!	34.21	Yes
62.25	–	62.25	No
36.54	Buuum!	29.02	Yes
34.21	Buuum!	36.54	Yes

Table 9.3 The normal pairing by the angle of inclination

α	Outcome	z	Result
11.32	–	11.32	No
28.77	Buuum!	28.77	Yes
29.02	Buuum!	29.02	Yes
30.78	Buuum!	30.78	Yes
34.21	Buuum!	34.21	Yes
36.54	Buuum!	36.54	Yes
62.25	–	62.25	No
79.56	–	79.56	No
88.23	–	88.23	No

Finally, compare this with the normal situation when the pairing is based on the angle of inclination (Table 9.3). This parametrization is based on a real physical circumstance establishing real conjunctive relation between the elements of the two families. There is maximal correlation; condition (b) is satisfied.

The upshot is that parameter λ in conditions (a) and (b) is not completely arbitrary[3]; not simply an "abstract" or conceived parametrization. It must have a

---

[3]I am grateful to Márton Gömöri for calling my attention to this issue.

physical realization, establishing a real conjunctive relationship between a_λ and A_λ, by which condition (b) expresses a real correlation between physical facts of L and physical facts of U.

With this strengthening all elements of semantic relationship become accommodated in the physical world.

9. A few important remarks are in order.

1. As we have seen, to be a meaning-carrier is not simply a matter of convention or definition or declaration. Semantics *is not an arbitrary assignment* of states of affairs of the world to linguistic elements of the theory.
2. It is pointless to talk about the meaning of an *isolated* formula of the theory. (Semantic holism) It is not only because of condition (a), but also because in condition (b) a big part of the axiomatic system can be involved.
3. It must be recognized that condition (b) is nothing but the necessary and sufficient condition for the theory (L, S) to be *true*. That is, the two conceptions *meaning and truth are completely intertwined.*
4. As semantics, in the above holistic sense, is a part and parcel of physical theory, no physical theory without semantics. Therefore, in case of empirical failure of a physical theory, semantics is one of the possible subjects of revision. In other words, semantics is as much hypothetical as any other part of the theory. For example, given that L is consistent, one can easily see that the following statements cannot hold true at the same time:

   (i) A refers to a
   (ii) $L \vdash A$
   (iii) a is not the case in U

   since (i) and (iii) would imply $L \vdash \neg A$. Therefore, observing that a is not the case we are not entitled to say that we observe that "$\neg A$". Simply because if a is not the case, then condition (b) fails, the whole semantics is lost. Therefore $\neg A$ does not carry meaning at all. That is to say, the state of affairs when a is not the case is something unexpressed, a brute phenomenon. This sheds light on the constitutive role of semantics, or rather, of the whole theory, in the sense of Reichenbach's "constitutive a priori" (Reichenbach 1965).

10. Thus, as we discussed in point **8**, condition (b) requires a real correlation between physical facts of two parts of the physical world, L and U. Combining this with the thesis of

(1) *the causal closeness of the physical world,* and
(2) *the principle of common cause,*[4]

---

[4]I mean the Reichenbachian thesis that no correlation without causation; every correlation is either due to a direct causal effect, or is brought about by a third factor, a so-called common cause (e.g. Reichenbach 1956; Hofer-Szabó et al. 2013).

one must conclude that both semantic relationship and the truth of the physical theory (consequently, our *knowledge*) must be brought about by the underlying causal processes of the physical world, going on in the common causal past of the two parts of the world L and U. This underlying process is what we normally call *learning through experience*. This is a very strong support of empiricism. No knowledge of the physical world is possible without experience. By the same token, no semantically meaningful talk about the physical world is possible without experience. There is no a priori meaning and there is no a priori truth.

In agreement with what was said in point **4**, it is worthwhile mentioning that the same is true with respect to the meta-mathematical theories. According to the physico-formalist claims, a formal system is a particular part of the physical world. Any theory describing a formal system, therefore, describes a part of the physical world; that is, it is a particular case of physical theory. There is no a priori meaning and truth in a meta-mathematical theory either.

11. It is a deep belief of many physicists that mathematics provides us the key ideas that are realized in the natural world. In Einstein's words:

> Our experience up to date justifies us in feeling sure that in Nature is actualized the ideal of mathematical simplicity. It is my conviction that pure mathematical construction enables us to discover the concepts and the laws connecting them which give us the key to the understanding of the phenomena of Nature. Experience can of course guide us in our choice of serviceable mathematical concepts; it cannot possibly be the source from which they are derived; experience of course remains the sole criterion of the serviceability of a mathematical construction for physics, but the truly creative principle resides in mathematics. In a certain sense, therefore, I hold it to be true that pure thought is competent to comprehend the real, as the ancients dreamed. (Einstein 1934, 167)

One might raise the question, if mathematics is only about the formal systems without meaning, how it is, then, possible that mathematical structures prove themselves to be so expressive in the physical applications. As Richard Feynman put it: "I find it quite amazing that it is possible to predict what will happen by mathematics, which is simply following rules which really have nothing to do with the original thing." (Feynman 1967, 171)

Let me start with mentioning that it is not mathematics *alone* by which the physicist can predict what will happen, but physical axioms and mathematics together. The physical axioms are determined by empirical facts. More exactly, the physicist, keeping, as long as possible, the logical and mathematical axioms fixed, *tunes* the physical axioms such that the theorems *derivable* from the unified system of logical, mathematical, and physical axioms be compatible with the empirical facts. Consequently, the employed logical and mathematical structures in themselves need not reflect anything about the real world in order to be useful.

Let me explain this with an analogy. You can experience a similar situation when you change the mouse driver on your computer (or just change the mouse settings): first you feel that the pointer movements ("derived theorems") generated by the new driver ("mathematics") according to your previously habituated hand movements ("physical axioms") do not faithfully reflect the geometry of your screen. Then, keeping the driver (and driver settings) fixed, you *tune* your hand movements – through typical "trial and error" learning – such that the generated pointer movements fit to the arrangement of your screen content.

Thus, there is no miraculous "preadaption" involved just because certain aspects of empirical reality "fit themselves into the forms provided by mathematics". This is simply a result of selections made by the physicist. Just as there is no preadaption at work when you successfully can install kitchen units obtained from a department store in your kitchen. The rules according to which the shelves, cupboards and doors can be combined show nothing about the actual geometry of your kitchen. But the final result is that the kitchen "fits itself" to the form of the whole set, as if through a kind of preadaption.

12. The constitutive role of formal systems in our physical knowledge by no means entitles us to say that there is a hypostatized *a priori* conceptual scheme in terms of which we grasp the experienced physical reality, and that this conceptual scheme generates analytic truths. For, what there is is anything but not conceptual: we only have the flesh and blood formal systems without any reference or meaning. Once an otherwise meaningless formula of a formal system is provided with meaning, in the sense of point **7**, it becomes true or false in a non-analytic sense.

Acknowledgements The research was partly supported by the (Hungarian) National Research, Development and Innovation Office, No. K100715 and No. K115593.

References

Ayer, Alfred J. 1952. *Language,truth and logic*. New York: Dover Publications.
Bell, Eric T. 1951. *Mathematics:Queen and servant of science*. New York: McGraw-Hill Book Company.
Colyvan, Mark. 2000. Conceptual contingency and abstract existence. *The Philosophical Quarterly* 50: 87–91.
Colyvan, Mark. 2004. Indispensability arguments in the philosophy of mathematics. *The Stanford encyclopedia of philosophy (Fall 2004 Edition)*, ed. Edward N. Zalta.
Crossley, J.N., C.J. Ash, J.C. Stillwell, N.H. Williams, and C.J. Brickhill. 1990. *What is mathematical logic?* New York: Dover Publications.
Curry, Haskell B. 1951. *Outlines of a formalist philosophy of mathematics*. Amsterdam: North-Holland.
David Deutsch, Artur Ekert, and Rossella Lupacchini. 2000. Machines, logic and quantum physics. *Bulletin of Symbolic Logic* 6: 265–283.
Einstein, Albert. 1934. On the method of theoretical physics. *Philosophy of Science* 1: 163–169.
Feynman, Richard. 1967. *The character of physical law*. Cambridge: MIT Press.
Field, Hartry. 1993. The conceptual contingency of mathematical objects. *Mind* 102: 285–299.
Hofer-Szabó, Gábor, Miklós Rédei, and László E. Szabó. 2013. *The principle of the common cause*. Cambridge: Cambridge University Press.
Miller, Kristie. 2012. Mathematical contingentism. *Erkenntnis* 77: 335–359.
Reichenbach, Hans. 1956. *The direction of time*. Berkeley: University of California Press.

Reichenbach, Hans. 1965. *The theory of relativity and a priori knowledge*. Berkeley: University of California Press.
Szabó, László E. 2003. Formal system as physical objects: A physicalist account of mathematical truth. *International Studies in the Philosophy of Science* 17: 117–125.
Szabó, László E. 2012. Mathematical facts in a physicalist ontology. *Parallel Processing Letters* 22: 1240009.

Chapter 10
Indeterminism, Gravitation, and Spacetime Theory

Samuel C. Fletcher

Abstract Contemporary discussions of physical determinism that engage with modern spacetime and gravitational theory have typically focused on the question of the *global* uniqueness of solutions for initial-value problems. In this chapter I investigate the violation of local uniqueness, found in examples like Norton's dome, which are not typically considered in light of spacetime theory. In particular, I construct initial trajectories for massive particles whose worldlines are not uniquely determined from initial data, both for a charged particle in special relativistic electromagnetism and for a freely falling particle in pure general relativity. I also show that the existence of such examples implies the violation of the Strong Energy Condition, and consider their implications for the interpretation of spacetime theory.

10.1 Introduction: Varieties of Indeterminism

Contemporary discussions of physical determinism that engage with modern spacetime and gravitational theory (e.g., Earman 1986, 2007) have typically focused on the question of the *global* uniqueness of solutions for two sorts of initial-value problems: that of the evolution of a field or a number of massive point particles in a fixed spacetime, or of four-dimensional spacetime itself from a three-dimensional "slice" within the context of general relativity. (I set the "indeterminism" sometimes associated with the Hole Argument aside. Cf. Brighouse 1994.) Paradigm examples of indeterminism in these contexts involve, respectively, the non-collision singularities of the "space invaders" scenario, where the worldliness of gravitating massive

---

Thanks to Dennis Dieks and the audience in Budapest for the Fifth Budapest-Krakow Research Group Workshop on May 23–24, 2016 for their helpful comments.

S.C. Fletcher (✉)
Department of Philosophy, University of Minnesota, Twin Cities, Minneapolis, MN, USA

Munich Center for Mathematical Philosophy, Ludwig Maximilian University of Munich, Munich, Germany
e-mail: scfletch@umn.edu

G. Hofer-Szabó, L. Wroński (eds.), *Making it Formally Explicit*, European Studies in Philosophy of Science 6, DOI 10.1007/978-3-319-55486-0_10

particles appear from spatial infinity after an arbitrary time, and the existence of non-unique maximal extensions, where a given "slice" can be evolved in one of many incompatible ways.

Here I shall investigate another sort of indeterminism, which, though studied in the literature on determinism in classical physics (Hoefer 2016, §4), is not typically considered in light of spacetime theory. In this sort of indeterminism, a localized, point-like physical system with equations of motion described by a set of differential equations has an initial-value problem with many solutions—typically uncountably infinitely many. This is a failure of *local* uniqueness of solutions. While systems exhibiting this property have been known since at least the nineteenth century (Fletcher 2012; van Strien 2014), a simple, concrete example due to Norton (2003, 2008) has recently captured philosophers' attention. In this example, a massive point particle begins at rest on the surface of a peculiarly shaped cylindrically symmetric rigid dome. The difference in height of a point on the dome from its apex, h, can be given as a function of the radial distance to that point from the apex along the surface of the dome, r:

$$h(r) = (2b^2/3g)r^{3/2}, \tag{10.1}$$

where g is the (constant) acceleration due to gravity at the surface of the Earth and b is a dimensional constant. If the dome is fixed rigidly to a flat surface in a uniform gravitational field yielding acceleration g, then the net force on the massive particle is the component of the total gravitational force tangential to the surface:

$$F_{\parallel} = mg \sin \theta = mg(dh/dr) = mb^2\sqrt{r}, \tag{10.2}$$

where θ is the angle between the horizontal and the radial tangent to the dome's surface at the particle's location.[1] Now, if the particle's initial position at $t = 0$ is the apex of the dome, then its initial value problem has as a solution

$$r(t) = \begin{cases} 0, & \text{if } t \leq T, \\ \frac{1}{144}(b[t - T])^4, & \text{if } t > T, \end{cases} \tag{10.3}$$

for *every* positive constant T. In other words, it is compatible with the Newtonian dynamics for the particle to fall down an arbitrary side of the dome after an arbitrary time resting at the top.

Mathematically, the local uniqueness to an initial value problem can be guaranteed only when further conditions hold of the equations of motion (Agarwal and Lakshmikantham 1993). In the case of Newtonian point particles, one such condition is that the force on the particle be *locally Lipschitz continuous* at the initial conditions (Arnol'd 1992, 36–38, 104–105). In general, a function $F(x) : D \to \mathbb{R}^m$,

---

[1]Note that since $0 \leq \theta \leq 90°$, $0 \leq dh/dr \leq 1$, hence $0 \leq r \leq g^2/b^4$ and $0 \leq h \leq 2b^2/3g^4$: the dome must have a finite height.

with $D \subseteq \mathbb{R}^n$, is *Lipschitz continuous on D* just when there is a constant $K > 0$ such that for all $x, y \in D$, $|F(x) - F(y)| \leq K|x - y|$. *F* is locally Lipschitz continuous at $x \in D$ when there is some neighborhood *U* of *x* on which it is Lipschitz continuous. It is easy to show that $F_{\|}$ in Equation 10.2 is not Lipschitz continuous at $r = 0$: $|F_{\|}|/|r|$ is not bounded above as $r \to 0$. Physically, this corresponds to an initial condition for the particle in which its net force (hence net acceleration) vanishes, but the derivative of that force in any spatial direction is infinite. This infinite derivative is necessary, but not sufficient, for nonuniqueness of the initial value problem, as the example of a ball rolling off a table illustrates.

The ensuing discussion of Norton's dome has mostly concentrated on furnishing various arguments against the dome system's legitimacy, either as an incomplete or incorrect application of Newtonian theory, or as an unphysical or otherwise improper idealization in that theory. (See Fletcher (2012) for an overview of many of these objections.) By contrast, Malament (2008) and Wilson (2009) have rejected the question as being too simplistic, for it assumes that there is a mathematically univocal, common conception of Newtonian mechanics. Elaborating on this idea, Fletcher (2012) argues for a plurality of closely related theories of Newtonian mechanics, some of which may have indeterministic models, partly on the grounds that indeterminism arises from mathematical features of the equations of motion for the models rather than from any identifiably common physical feature thereof.

The goal of this paper is to expand on this thesis by suggesting that the sort of indeterminism given by multiple solutions to a test particle's initial value problem depends neither on the physics in question being non-relativistic, nor on the specification of a particular form of forced motion. Its aspiration, in other words, is the construction of a certain relativistic spacetime, some of whose timelike geodesics passing through a point are not uniquely determined by the specification of their tangent vector there. In such a spacetime, the woldlines of free massive test particles modeled by these geodesics exhibit the sort of indeterminism displayed by Norton's dome. This indeterminism also combines in a novel way some of the features of both of the two sorts of indeterminism described at the beginning of this section: on the one hand, it applies to particular sets of worldliness of massive test particles without varying spacetime structure, and on the other, it does so without introducing new matter entering the universe from spatial infinity.

To arrive at this example, I will attempt to make two successive modifications to an electrostatic example of Fletcher (2012), described in Sect. 10.2.1, which displays the same sort of indeterminism as the dome system. First, in Sect. 10.2.2, I adapt it from Newtonian to special relativistic physics. The second modification, in Sect. 10.2.3, attempts to replicate this sort of indeterminism with the unforced, or geodesic, motion of a test particle in a general relativistic spacetime.

These examples raise the question about the sorts of spacetimes, relativistic or otherwise, which admit of non-unique solutions to the geodesic equation for some initial conditions. I suggest a partial answer in Sect. 10.3, arguing that any spacetime exhibiting this sort of indeterminism must at some point violate the strong energy condition (SEC), which heuristically can be understood as the statement that the effects of gravity are locally attractive (i.e., causal geodesics tend to converge). In other words, repulsion, whether through forced or natural motion, is necessary for this sort of local indeterminism. This immediately implies that this sort of indeterminism cannot manifest in either relativistic vacuum spacetimes or through pure gravity in Newtonian spacetimes.

Other sorts of questions raised by these examples, discussed in the concluding Sect. 10.4, are of an interpretive nature. In addition to the examples providing more evidence for a pluralistic understanding of relativity theory, they also unsettle the usual, though tacit, assumption that the specification of a relativistic spacetime is a complete determination of all the events of a model world. One option is to augment spacetime structure to fix the actual worldlines of test particles; another, more ambitious but less developed option is to take the concept of the test particle and the worldline more seriously as idealizations, delimiting their range of fruitful application more precisely. I provide some, though not entirely conclusive, reasons that the latter option is to be preferred.

10.2 Indeterminism Through Forced and Unforced Motion

In the sections below concerning relativistic spacetimes, I use the abstract index notation (Malament 2012, §1.4; Wald 1984, §2.4) according to which lowercase superscript (resp. subscript) roman letters $(a, b, c, \dots)$ on a symbol representing a tensor denote the label of the vector (resp. covector) space(s) in which the tensor lives. As an example of this notation: those sections will consider relativistic spacetimes (M, g_{ab}), where M denotes a connected, paracompact, four-dimensional real manifold and g_{ab} denotes a Lorentz metric of signature $(1, 3)$ on this manifold.

Note as well that Sect. 10.2.1 is based on Fletcher (2012, §3.4).

10.2.1 Indeterminism of Forced Motion in Newtonian Spacetime

Consider the following spherically symmetric electric charge distribution in a Newtonian spacetime:

$$\rho(r) = \begin{cases} 5C\epsilon_0/2\sqrt{r}, & \text{if } 0 < r \leq R, \\ 0, & \text{if } r = 0 \text{ or } r > R, \end{cases} \tag{10.4}$$

where r is the radial distance from the center of symmetry, ϵ_0 is the permittivity of free space, and C is a dimensional constant. Even though $\lim_{r \to 0} \rho(r) = \infty$, the total charge Q is finite:

$$Q = \int \rho dV = 4\pi C\epsilon_0 R^{5/2}. \tag{10.5}$$

By Gauss's Law, the radial component of the electric field for $r \leq R$ is

$$E_r(r) = \frac{1}{4\pi\epsilon_0 r^2} \int_{B_r} \rho dV = C\sqrt{r}, \tag{10.6}$$

where B_r is the ball of radius r. A test particle with charge q starting from rest at a radial distance $r \leq R$ from the center of the charge distribution experiences a Coulomb force

$$F_r(r) = qE_r(r) = qC\sqrt{r}. \tag{10.7}$$

Thus, if the particle is initially at rest at the origin, it has uncountably many solutions $r(t)$ to its equation of motion exactly in analogy with Equation 10.3.

10.2.2 Indeterminism of Forced Motion in Special Relativity

For convenience, choose units so that the numerical value of the speed of light c is 1, as is the permittivity of free space ϵ_0. Consider Minkowski spacetime $(\mathbb{R}^4, \eta_{ab})$ along with a constant, unit timelike vector field t^a whose geodesic congruence represents the worldlines of a family of inertial observers. In order to reproduce an analog of the example in Sect. 10.2.1, pick (the image of) one such worldline γ as the axis of symmetry for a charge distribution

$$\rho = \begin{cases} 5C/2\sqrt{r}, & \text{if } 0 < r \leq R, \\ 0, & \text{if } r = 0 \text{ or } r > R, \end{cases} \tag{10.8}$$

where, much as before, C is a dimensional constant and $r_{|p}$ is the distance from $p \in M$ to (the image of) γ along the spacelike geodesic orthogonal to it. Now, given any observer with four-velocity ξ^a at a point, they can reconstruct the Faraday tensor F_{ab} and the charge-current density J^a according to

$$F_{ab} = E_a\xi_b - \xi_a E_b + \varepsilon_{abcd}\xi^c B^d, \tag{10.9}$$

$$J^a = \rho\xi^a + j^a, \tag{10.10}$$

where E^a is their observed electric field, B^a is their observed magnetic field, j^a is their observed current density, and ε_{abcd} is a volume form (Malament 2012, p. 154). Thus we may assign, for the family of inertial observers determined by t^a,

$$E^a = \begin{cases} C\sqrt{r}(r^a), & \text{if } 0 \leq r \leq R, \\ (CR^{5/2}/r^2)r^a, & \text{if } r > R, \end{cases} \tag{10.11}$$

$$B^a = \mathbf{0}, \tag{10.12}$$

$$j^a = \mathbf{0}, \tag{10.13}$$

where r^a is the radial spatial vector field from γ. This yields that

$$F_{ab} = \begin{cases} C\sqrt{r}(r_a t_b - t_a r_b), & \text{if } 0 \leq r \leq R, \\ (CR^{5/2}/r^2)(r_a t_b - t_a r_b), & \text{if } r > R, \end{cases} \tag{10.14}$$

$$J^a = \begin{cases} (5C/2\sqrt{r})t^a, & \text{if } 0 < r \leq R, \\ \mathbf{0}, & \text{if } r = 0 \text{ or } r > R. \end{cases} \tag{10.15}$$

To verify that these are indeed solutions of Maxwell's equations,

$$\nabla_{[a}F_{bc]} = \mathbf{0}, \tag{10.16}$$

$$\nabla_a F^{ab} = J^b, \tag{10.17}$$

note first that one can define

$$A_a = \begin{cases} (2Cr^{3/2}/3)t_a, & \text{if } 0 \leq r \leq R, \\ -(CR^{5/2}/r)t_a, & \text{if } r > R, \end{cases} \tag{10.18}$$

so that

$$F_{ab} = \nabla_a A_b - \nabla_b A_a. \tag{10.19}$$

Substitution of Equation 10.14 into the left-hand side of Equation 10.16 then yields the desired result.

For Equation 10.17, substitution of Equation 10.14 and application of the product rule gives, for $0 \leq r \leq R$,

$$\begin{aligned} \nabla_a(C\sqrt{r}[r^a t^b - t^a r^b]) &= (C/2\sqrt{r})r_a(r^a t^b - t^a r^b) \\ &\quad + C\sqrt{r}([\nabla_a r^a]t^b + r^a \nabla_a t^b - [\nabla_a t^a]r^b - t^a \nabla_a r^b), \\ &= (C/2\sqrt{r})t^b + C\sqrt{r}([2/r]t^b + 0 + 0 + 0), \\ &= (5C/2\sqrt{r})t^b, \end{aligned}$$

and for $r > R$,

$$\nabla_a(CR^{5/2}/r^2[r^a t^b - t^a r^b]) = (-2CR^{5/2}/r^3)r_a(r^a t^b - t^a r^b)$$
$$+ (CR^{5/2}/r^2)([\nabla_a r^a]t^b + r^a \nabla_a t^b - [\nabla_a t^a]r^b - t^a \nabla_a r^b),$$
$$= -(2CR^{5/2}/r^3)t^b + (CR^{5/2}/r^2)([2/r]t^b + 0 + 0 + 0),$$
$$= 0,$$

which matches Equation 10.15 as required.

Lastly, consider a test particle of charge q and mass m initially co-moving with the charge distribution. If its initial position p is within the charge distribution (i.e., $r_{|p} < R$), then it experiences a Lorentz force $qF^a{}_b t^b = qC\sqrt{r}r^a$, which is not Lipschitz continuous at $r = 0$. Thus, if the particle's initial location p satisfies $r_{|p} = 0$, then there are infinitely many worldlines which result as solutions to its equation of motion: $d^2r/d\tau^2 = (q/m)C\sqrt{r}$, where τ is the proper time along the particle's worldline. Note that these solutions are only locally in analogy with those of Equation 10.3, for the latter in principle can result in unbounded velocities. In contrast, the relativistic equation of motion concerns proper time τ, not the coordinate time t, but for sufficiently small relative velocities with the inertial observers posited at the beginning of this section, they approximate each other arbitrarily well.

10.2.3 Indeterminism of Natural Motion in General Relativity

As with virtually all previous examples of locally nonunique solutions to a differential equation of motion, the one described above in Minkowski spacetime with a carefully chosen distribution of matter renders a force on a test particle that is non-Lipschitz at certain points. The same failure of uniqueness, however, can also arise in general relativity from the geodesic motion of a test particle—that is, from gravity alone.

Consider any relativistic spacetime (M, g_{ab}) with Levi-Civita connection ∇.[2] The (proper) acceleration for a test particle whose wordline has tangent vector ξ^a can be expressed using a locally flat derivative operator ∂ as

$$\xi^b \nabla_b \xi^a = \xi^b \partial_b \xi^a - \xi^b \xi^m C^a{}_{bm}, \tag{10.20}$$

---

[2]From the present perspective, the stress-energy tensor T_{ab} is not an independent object for a relativistic spacetime once the metric has been specified, since the Riemann tensor $R^a{}_{bcd}$ associated with the Levi-Civita connection determines the Ricci tensor R_{ab} and curvature scalar R, which in turn determine T_{ab} through Einstein's equation, $T_{ab} = (1/8\pi)(R_{ab} - (1/2)Rg_{ab})$.

where $C^a{}_{bm}$ is the connection tensor between ∇ and ∂ whose components in a particular coordinate basis are the Christoffel symbols (Malament 2012, Prop. 1.7.3; Wald 1984, p. 34). The connection tensor in turn can be expressed using the metric (Malament 2012, eq. 1.9.6; Wald 1984, eq. 3.1.28):

$$C^a{}_{bm} = \frac{1}{2}g^{an}(\partial_n g_{bm} - \partial_b g_{nm} - \partial_m g_{nb}).$$ (10.21)

Thus, by substituting Equation 10.21, we can rewrite Equation 10.20 as

$$\xi^b \nabla_b \xi^a = \xi^b \partial_b \xi^a - \frac{1}{2}\xi^b \xi^m g^{an}(\partial_n g_{bm} - \partial_b g_{nm} - \partial_m g_{nb}).$$ (10.22)

In coordinates adapted to the flat connection ∂, the right-hand side expresses the "force" on the test particle (up to a factor of the mass of the particle). Now, the *true* force on a test particle is not a coordinate-dependent quantity. But what matters here is that Equation 10.22 has the same form as Newton's second law, considered as a differential equation. Thus, through an appropriate choice of g_{ab}, one can design the same non-uniqueness to its solutions for appropriate choices of initial conditions.

Implementing this feature requires the first derivatives of the spacetime metric with respect to ∂ to be non-Lipschitz continuous at some point, so the metric cannot be everywhere smooth. While smoothness is often demanded of the spacetime metric, its full strength is not required to formulate general relativity adequately. As the example below will show, allowing it to be C^1 on a one-dimensional line still enables one to define all the needed geometric objects of the theory, while still allowing for non-unique solutions to the initial value problem for some geodesics.

In analogy with the examples of this indeterminism already considered, I restrict attention to static, spherically symmetric spacetimes on $\mathbb{R}^4$, whose metrics and inverse metrics take on the general form (Wald 1984, eq. 6.1.5)

$$g_{ab} = e^{2\nu} t_a t_b - e^{2\lambda}(r_a r_b + r^2(\theta_a \theta_b + \sin^2 \theta \phi_a \phi_b)),$$ (10.23)

$$g^{ab} = e^{-2\nu} t^a t^b - e^{-2\lambda}(r^a r^b + r^{-2}(\theta^a \theta^b + \csc^2 \theta \phi^a \phi^b)),$$ (10.24)

where I have used the abbreviations $x_a = \partial_a x$ and $x^a = (\partial/\partial x)$ for $x \in \{t, r, \theta, \phi\}$, spherical coordinate fields well-adapted to the symmetries of the spacetime, and where ν and λ depend only on r. Further, I restrict attention to test particles whose initial four-velocity is t^a. The goal is to find a metric whose derivatives yield an acceleration field proportional to $\sqrt{r}$ in the right-hand side of Equation 10.22.

To analyze this problem, first consider the acceleration of *any* test particle whose initial four-velocity is t^a, not necessarily one undergoing geodesic motion. Substitution of Equation 10.23 into Equation 10.22 gives that

$$t^b \nabla_b t^a = t^b \partial_b t^a - \frac{1}{2}t^b t^m g^{an}(\partial_n g_{bm} - \partial_b g_{nm} - \partial_m g_{nb})$$

$$= g^{an} t^b \partial_b(t^m g_{nm}) - \frac{1}{2}g^{an}\partial_n(t^b t^m g_{bm}).$$ (10.25)

Since v and λ depend only on r, the product rule for differentiation then yields that

$$\partial_b(t^m g_{nm}) = 2v'e^{2v}t_n r_b, \tag{10.26}$$

$$\partial_n(t^b t^m g_{bm}) = 2v'e^{2v} r_n, \tag{10.27}$$

where $v' = dv/dr$. Combining these with Equations 10.24 and 10.25,

$$t^b \nabla_b t^a = g^{an} t^b (2v'e^{2v} t_n r_b) - \frac{1}{2} g^{an}(2v'e^{2v} r_n) = v'e^{2v-2\lambda} r^a. \tag{10.28}$$

There are many possible choices of v and λ which will yield a function on the right-hand side that is not everywhere Lipschitz continuous. For simplicity, I will examine the case of

$$v = \lambda = A + (2B^2/3)r^{3/2}, \tag{10.29}$$

where A and B are constants. For this choice, the spacetime is conformally equivalent to Minkowski spacetime, i.e.,

$$g_{ab} = e^{2A+(4B^2/3)r^{3/2}} \eta_{ab}, \tag{10.30}$$

and

$$t^b \nabla_b t^a = B^2 \sqrt{r} r^a. \tag{10.31}$$

Thus a test particle initially at any point p with $r_{|p} = 0$ and four-velocity t^a undergoes geodesic motion if all the points of its wordline satisfy $r = 0$. But this is not the only geodesic passing through p with tangent vector t^a, for there are infinitely many solutions to the equation $dr^2/d\tau^2 = B^2\sqrt{r}$ with initial condition $r = 0$. Just as in the previous subsection, these solutions are only locally analogous to those of Equation 10.3, due to the relativistically necessary proper time. But the way that the "gravitational field"—really, just spacetime curvature—emulates the Maxwellian field in the role as the source of spontaneous "acceleration" is similar.

The new metric given by Equation 10.30 is clearly smooth everywhere except on the line of $r = 0$, where it is merely C^1. Thus the Riemann tensor $R^a{}_{bcd}$ associated with the Levi-Civita connection, which is necessary to define the Ricci tensor R_{ab} in Einstein's equation, is well-defined in the usual way everywhere except on the line of $r = 0$.[3] Here, it is most natural to set $(R^a{}_{bcd})|_{r=0} = \mathbf{0}$, for observers with worldlines on $r = 0$ measure the metric to be Minkowskian (up to an immaterial

---

[3]One may apply the formulas of Wald (1984, p. 446) to calculate these explicitly, although doing so does not seem to give any obvious insight into the nature of the indeterminism this spacetime exhibits.

constant factor). The resulting curvature, while well-defined everywhere, does not vary continuously, much as the electric charge density did not in the previous sections.

10.3 Indeterminism and the Strong Energy Condition

The matter distributions and spacetimes considered above are unusual. Are there *general* criteria for deciding when a worldline of a test particle is not deterministic? A failure of smoothness in the metric or matter distribution is certainly necessary, but I suspect more can be said, in particular in connection with the SEC. Following Curiel (2017, §2.1), one can distinguish (at least) between a geometric version of the SEC, which states that

$$R_{ab}\xi^a\xi^b \geq 0 \tag{10.32}$$

for any timelike vector ξ^a, and a physical version of the SEC, which states that

$$(T_{ab} - \frac{1}{2}Tg_{ab})\xi^a\xi^b \geq 0 \tag{10.33}$$

for any timelike vector ξ^a. The physical version is implied by the geometric version under the assumption of Einstein's equation, but the geometrical version has a natural interpretation, namely that timelike geodesics locally converge. Below I sketch an argument that a necessary condition for the nonuniqueness of a solution to a test particle's initial value problem at some point p is a violation of the SEC in a neighborhood of p.

Consider a spacetime with a C^1 metric in which the initial value problem for the geodesic motion of a point test particle at $p \in M$ has more than one solution, and let τ^a be the tangent vector field of these solutions, extended if necessary to a neighborhood around them, and pick a triple of vector fields x^a, y^a, z^a defined on the first solution that are orthogonal to and Lie derived by τ^a. Together, the four form a geodesic reference frame which we may use to define a coordinate system in a neighborhood of the initial point p, with the first local solution lying at the (x, y, z) coordinate origin. We may equip the other solution with a tetrad as well that is Lie derived by (a local extension to a neighborhood of) its tangent vector τ^a, and in particular has one component as the connecting field representing the relative acceleration between the two solutions. The contrapositive to the Peano uniqueness theorem (Agarwal and Lakshmikantham 1993, §3.3) states that this must be on average increasing. But it is well known that the *average* radial acceleration amongst the three independent spatial directions is proportional to the Raychaudhuri scalar:

$$ARA = -\frac{1}{3}R_{ab}\tau^a\tau^b, \tag{10.34}$$

which is the quantity bounded by the SEC. But ARA here is positive, so the geometric SEC is violated. Thus:

Proposition *The geometric SEC is violated in the neighborhood of a point of a spacetime where the initial value problem for the geodesic equation does not have a unique solution.*

In other words, gravitation must become "repulsive" as this point is approached. Moreover, since Einstein's equation sets $R_{ab} = T_{ab} - \frac{1}{2}Tg_{ab}$, where T_{ab} is the stress-energy tensor and T is its trace, this implies the right-hand, contracted with the timelike vectors $\xi^a\xi^b$ becomes negative as this point is approached. And *this* in turn implies that the physical SEC is violated.

One of the interesting aspects of this argument is that substantive assumptions particular to relativity theory only entered at the end, in the invocation of Einstein's equation, and only there to connect the constraint, which is really on the Raychaudhuri scalar, with a well-known energy condition. Thus, one can apply a similar argument to other spacetime theories. In geometrized Newtonian gravitation, for example, the Raychaudhuri scalar is equated with the mass density through the geometrized version of Poisson's equation (Malament 2012, p. 269). So the analog of the SEC in that theory is just the condition that mass be non-negative. Insofar as this is a central, not auxiliary, assumption of Newtonian gravitation, the above argument then yields that the sort of indeterminism considered in this paper is not possible with Newtonian gravitation alone. The reason for this comes again in the interpretation of the ARA: Newtonian gravitation is always attractive, never repulsive, while repulsivity is a necessary condition for indeterminism.

10.4 Discussion and Conclusions

Perhaps the first question raised by the foregoing examples is whether they should be excluded from legitimacy. It is easy to do so by reaffirming the demand for only smooth spacetime structures, for example. But this would rule out the discontinuous matter distributions and non-smooth spacetime metrics used in modeling stars (as objects with compact boundaries), shock waves (Israel 1960; Smoller and Temple 1997), and colliding black holes (Merritt and Milosavljević 2005). Demanding that the SEC holds seems in conflict with the use of successful models, especially in cosmology, that violate it persistently (Curiel 2017). One could also restrict attention only to "physically reasonable" distributions of known or pedestrian matter, but the grounds for doing so may be questionable (Manchak 2011).

Perhaps it is best, as Fletcher (2012) advocates with classical (non-relativistic) mechanics, to decline to make a decision and instead embrace the idea that there are many different versions of the theory of general relativity, some more delimited and other less so, some in which the above sorts of indeterminism are allowed and others in which they are eliminated. For, by allowing different versions of our theories to coexist, we can gain more insight into how their different parts fit together. In particular, further investigation of non-smooth spacetimes might perhaps allow for a new kind of response to the singularity theorems (Curiel and Bokulich 2012) and other results about non-extendibility, in two ways: first, those results tend to assume that the collection of spacetimes that can be extended into must be smooth—indeed, on that mark the general relativistic spacetime I considered above would have to have its origin excised; and second, they also send to assume some version of the SEC (Curiel 2017), whose violation, we saw in the previous section, is necessary for the failure of determinism. Is there some well-motivated way of performing non-smooth extensions to spacetimes to avoid singularities?

The other significant question raised by this investigation, to my mind, is about the exact role that the idealization of the point particle and its worldline are supposed to play in modern spacetime theory. The use and interpretation of some of the most basic concepts of the theory, such as the classification of tangent vectors into timelike, null, and spacelike, and the definitions of singularities and causality conditions, invoke the notion of a worldline of a point particle. Some approaches to the foundations of relativity theory even take them as primitive. But in the above models, they are not determined by the usual sorts of data provided. One cannot maintain, in these models, that the events of the spacetime manifold encode all the goings on, here and elsewhere, past, present, and future, and at once hold that some particles' histories are not so determined. It reveals a tension between the "principal" nature of general relativity's foundation—interpretive principles about worldlines seemingly grounded in firm evidence—with the "constructive" nature of the particular matter fields we might wish to model upon it (Einstein 1954).

One response to all this is to add structure, a further specification of the worldlines really occupied by particles. In some sense, the branching spacetimes framework (Placek 2014) does something like this, although not quite in this context: again, it is not the spacetime structure that is indeterministic on the example considered above, but the test particles within them. The difficulty with this response is that it gives up on a sort of reductionism that, while not mandated by the spacetime picture, is friendly with it: namely, the view that all matter, particles included, are fields on spacetime. Once one fixes those fields, all is determined. Reifying these indeterministic examples through an addition to the theory would seemingly abandon that.

Another response, and the one I tentatively prefer, is a careful reevaluation of the notion of a test particle and the role it plays in the foundations of spacetime theory. On this view, it is an extremely convenient and expedient idealization, but one whose limits need to be more clearly addressed. As a sort of infinite idealization, test particles cannot share or well-approximate all the properties or features of their de-idealized, extended, internally interacting field-theoretic counterparts. Just as infinite systems in statistical mechanics have features that no finite systems share, so too do test point particles with matter fields. This is not to say they should be extricated from the theory—far from it—but that further work is needed to understand their explanatory role, and the limits of their applicability. This paper has not attempted an answer to this question, but my hope is that it will stir the spirits of others to it.

References

Agarwal, R.P., and V. Lakshmikantham. 1993. *Uniqueness and nonuniqueness criteria for ordinary differential equations*. Singapore: World Scientific.

Arnol'd, Vladimir. 1992. *Ordinary differential equations* (Translation: Cooke, R.). Berlin: Springer.

Brighouse, Carolyn. 1994. Spacetime and holes. *PSA: Proceedings of the Biennial Meeting of the Philosophy of Science Association, 1994*, 117–125.

Curiel, Erik and Peter Bokulich. 2012. Singularities and black holes. In *The Stanford encyclopedia of philosophy (Fall 2012 Edition)*, ed. Edward N. Zalta. https://plato.stanford.edu/archives/fall2012/entries/spacetime-singularities/.

Curiel, Erik. 2017. A primer on energy conditions. In *Towards a theory of spacetime theories*, ed. Dennis Lehmkuhl, Gregor Schiemann, and Erhard Scholz, 43–104. Basel: Birkhäuser.

Earman, John. 1986. *A primer on determinism*. Dordrecht: D. Reidel.

Earman, John. 2007. Aspects of determinism in modern physics. In *Handbook of the philosophy of science, vol. 2: Philosophy of physics*, ed. John Earman and Jeremy Butterfield, 1369–1434. Amsterdam: Elsevier.

Einstein, Albert. 1954. What is the theory of relativity? In *Ideas and opinions*, ed. Albert Einstein, 227–232. New York: Crown.

Fletcher, Samuel C. 2012. What counts as a Newtonian system? The view from Norton's Dome. *European Journal for Philosophy of Science* 2(3): 275–297.

Hoefer, Carl. 2016. Causal Determinism. In *The Stanford encyclopedia of philosophy (Spring 2016 Edition)*, ed. Edward N. Zalta. http://plato.stanford.edu/archives/spr2016/entries/determinism-causal/.

Israel, W. 1960. Relativistic theory of shock waves. *Proceedings of the Royal Society of London: Series A, Mathematical and Physical Sciences* 259: 129–143.

Malament, David B. 2008. Norton's slippery slope. *Philosophy of Science* 75: 799–816.

Malament, David B. 2012. *Topics in the foundations of general relativity and newtonian gravitation theory*. Chicago: University of Chicago Press.

Manchak, John. 2011. What is a physically reasonable space-time? *Philosophy of Science* 78(3): 410–420.

Merritt, David and Miloš Milosavljević. 2005. Massive black hole binary evolution. *Living Reviews in Relativity* 8: 8.

Norton, John D. 2003. Causation as folk science. *Philosophers' Imprint* 3(4): 1–22.

Norton, John D. 2008. The dome: An unexpectedly simple failure of determinism. *Philosophy of Science* 75: 786–798.

Placek, Tomasz. 2014. Branching for general relativists. In *Nuel Belnap on indeterminism and free action*, ed. Thomas Müller, 191–221. London: Springer.

Smoller, J.A., and J.B. Temple. 1997. Shock-waves in general relativity. *Contemporary Mathematics* 208: 301–312.

van Strien, Marij. 2014. The Norton dome and the nineteenth century foundations of determinism. *Journal for General Philosophy of Science* 45(1): 167–185.

Wald, Robert M. 1984. *General relativity*. Chicago: University of Chicago Press.

Wilson, Mark. 2009. Determinism and the mystery of the missing physics. *British Journal for the Philosophy of Science* 60: 173–193.

Chapter 11
Non-uniquely Extendible Maximal Globally Hyperbolic Spacetimes in Classical General Relativity: A Philosophical Survey

Juliusz Doboszewski

Abstract I discuss philosophical questions raised by existence of extendible maximal globally hyperbolic spacetimes with non-unique extensions. These spacetimes represent a form of indeterminism in classical general relativity: given fixed initial data and equations of the theory, there exists more than one possible way a spacetime could be. Even though such spacetimes have been investigated in the physics literature for quite some time, a philosophical discussion of their importance is missing. Here I explore their relevance for the notion of physical equivalence, distinction between physically reasonable spacetimes and physically unreasonable ones, relation between determinism and singular spacetimes, connections between some forms of indeterminism and existence of time machines, and question whether cosmic censorship can be understood as expressing determinism of general relativity.

11.1 Introduction

I will discuss a few examples of extendible maximal globally hyperbolic spacetimes in classical general relativity (GR), and their bearing on the question whether general theory of relativity is deterministic. First, I provide overview of relevant background material (concerning notions of determinism and indeterminism, relativistic spacetimes and initial value formulation of classical general relativity). After that, I briefly present a few classess of spacetimes sharing an interesting property: in the initial value formulation, the nicely behaving maximal globally hyperbolic region of spacetime allows more than one continuation, in the sense that there exists more than one (up to isometry) inextendible spacetime into which this region can be isometrically embedded. Under some natural interpretation of what does it mean for a theory to be indeterministic, existence of such spacetimes shows that classical general relativity is indeterministic. But significance of such spacetimes

J. Doboszewski (✉)
Institute of Philosophy, Jagiellonian University, 52 Grodzka, 31-044 Kraków, Poland
e-mail: jdoboszewski@gmail.com

© Springer International Publishing AG 2017   193
G. Hofer-Szabó, L. Wroński (eds.), *Making it Formally Explicit*, European Studies
in Philosophy of Science 6, DOI 10.1007/978-3-319-55486-0_11

is not limited to question of determinism, which I will demonstrate by connecting these spacetimes to questions of physical equivalence in classical GR, conditions for distinguishing physically reasonable models from physically unreasonable ones, time machines, and interpretation of the cosmic censorship hypothesis.

11.1.1 Defining Determinism

Various non-equivalent precisifications of determinism have been offered in the literature. Often these characterizations invoke certain features of the physical situation (such as the existence of the set of moments of time, represented by the set of real numbers $\mathbb{R}$ with its natural linear ordering), features which, in the context of classical GR, do not hold in all of the solutions. One should thus be careful in how one spells out the notion of determinism, it may or may be not applicable to classical GR.

For example, Werndl (2015) provides the following intuition:

Determinism reigns when the state of the system at one time fixes the past and future evolution of the system.

And Wüthrich (2011), following Earman (1986), gives two definitions:

Definition 1 (Determinism for worlds) A world $W \in \mathcal{W}$ is deterministic if and only if for any $W' \in \mathcal{W}$, if W and W' agree at any time, then they agree for all times.

Definition 2 (Determinism for theories) A 'theory T is deterministic just in case, given the state description $s(t_1)$ at any time t_1, the state description $s(t_2)$ at any other time t_2 is deducible [in principle] from T.'

Determinism for theories is sometimes called Laplacian determinism, and spelled out in terms of models: a theory is deterministic iff every two models which agree at some t have to agree at all t'.

The trouble with such concepts is that in case of GR spacetime is a dynamical entity, and (prima facie) there are models of the theory which do not have anything like moments of time (Earman (2007) discussess a few concerns related to existence of acausal spacetimes in the context of Laplacian determinism). Fortunately, concept of Laplacian determinism can be spelled out in a way which is more friendly to pecularities of classical GR. One way to do so has been provided (in the context of debates surrounding the Hole Argument) by Butterfield (1989), and seems to be the most sophisticated version of Laplacian determinism available. According to Butterfield,

a theory with models $\langle M, O_i \rangle$ is **S**-deterministic, where **S** is a kind of region that occurs in manifolds of the kind occurring in the models, iff: given any two models $\langle M, O_i \rangle$ and $\langle M', O_i' \rangle$ containing regions S, S' of kind **S** respectively, and any diffeomorphism α from S onto S': if $\alpha^*(O_i) = O_i'$ on $\alpha(S) = S'$, then: there is an isomorphism β from M onto M' that sends S to S', i.e. $\beta^*(O_i) = O_i'$ throughout M and $\beta(S) = S'$.

Note that this definition does not assume anything like moments of time found in other explications of determinism. However, under this definition worry about existence of acausal spacetimes does not disappear entirely.

Some choices trivialize the question: if **S** is required to be a Cauchy surface of a spacetime in question, then (by Choquet-Bruhat and Geroch theorem I discuss in Sect. 11.1.4) uniqueness is ensured by fiat; but if **S** is just, say, some achronal submanifold of spacetime, uniqueness fails almost certainly. Either way question of determinism becomes trivial: choice of region **S** implies the answer to the question of determinism. Is there a choice of **S** which avoids this triviality? Such a choice would *de facto* implicitly assume that spacetimes which do not contain regions of type **S** (say, spacetimes violating some causality condition) are not physically reasonable, at least in the minimal sense that they are not relevant for the question of determinism: are neither examples of determinism nor indeterminism. One option, to which, I believe, insufficient attention has been paid by philosophers of science, is taking maximal globally hyperbolic development of an initial data set as **S** (I explain the terminology in Sect. 11.1.2). This choice has practical advantage: in a natural way it connects to the initial value problem in classical GR, and is what has been actually used by mathematical physicists. A consequence of that, however, is that spacetimes containing various kinds of holes which "mutilate" MGHD (i.e. spacetimes which have globally hyperbolic regions, but which do not contain maximal globally hyperbolic region as a subset) are excluded *by fiat*, in manner similar to Earman's take on the "dirty open secret" of Earman (1995).

Another subtle aspect is making precise what transformations are included as isomorphisms β. In most of what follows I will assume that an isometry (that is, pullback of the spacetime metric g_{ab} by a diffeomorphism) which preserves temporal orientation is the appropriate choice for β. But, as I discuss in Sect. 11.3.2, there are other possible choices one could make here.

Suppose (in case of relativistic spacetime) that spacetime is temporally orientable and that the orientation has been fixed. I will say that spacetime $\langle M, g \rangle$ is futuristically indeterministic if there are two spacetimes $\langle M', g' \rangle$ and $\langle M'', g'' \rangle$ which extend $\langle M, g \rangle$ in such a way that new regions of $\langle M', g' \rangle$ and $\langle M'', g'' \rangle$ are to the future of $\langle M, g \rangle$, and $\langle M', g' \rangle$ and $\langle M'', g'' \rangle$ are not isometric. Similarly, spacetime could be past indeterministic, or it could be indeterministic "in both ways". Interestingly, as I will show in what follows, classical GR provides us examples of all three types of indeterminism.

Another important distinction concerns indeterminism of the theories and determinism of the world. These two are very different. One can imagine that the best theory of the world is deterministic (in the sense that all models of the theory are deterministic), but the world itself is indeterministic, or the other way round—this could happen, for instance, when theory is incomplete. Indeed, we will argue that one can see extendible maximal globally hyperbolic spacetimes as showing that the theory of general relativity is indeterministic, in the sense that there are sectors of solutions in which general relativity has indeterministic solutions; but, at the same time, it seems that models most useful for the description of our world (FRLW spacetimes) are (at least in the classical theory) inextendible beyond the maximal globally hyperbolic region, which in turn strongly suggests that the world itself is deterministic, at least in the sense that it is (at the energy scales in which classical general relativity is uncontroversially applicable to our world, i.e. without taking into the account energy scales at which quantum effects are expected to play some role) best represented by a deterministic model.

11.1.2 Relativistic Spacetimes

A brief reminder on the structure of relativistic spacetimes is in order. A general relativistic spacetime is a pair $\langle M, g_{ab} \rangle$, where M is a smooth, second-countable, paracompact and Hausdorff manifold, and g_{ab} is a non-degenerate Lorentz-signature metric. In what follows I assume that spacetime satisfies Einstein's field equations, and for most of the time restrict our attention to vacuum solutions. I will follow the standard notation of Malament (2012).

Spacetime $\langle M, g_{ab} \rangle$ is globally hyperbolic iff there exists achronal subset without edge S which is intersected exactly once by every inextendible timelike curve (or, equivalently, S such that its domain of dependence is the whole spacetime, $D(S) = M$). Spacetime $\langle M, g_{ab} \rangle$ is extendible iff there exists spacetime $\langle M', g'_{ab} \rangle$ and a smooth function $f : M \mapsto M'$ which is a bijection onto its image, such that $f(M) \subsetneq M'$ and $g'_{ab} \upharpoonright_{f(M)} = f^*(g_{ab})$; otherwise it is inextendible. Spacetime is (timelike, null) geodesically complete iff generalized affine parameter of any (timelike, null) geodesics takes arbitrary values in $\mathbb{R}$.

Spacetime $\langle M', g'_{ab}, \Lambda \rangle$ is an extension of $\langle M, g_{ab} \rangle$, if:

1. there exists a function $\Lambda : M \mapsto M'$ which is an embedding of M in $\Lambda(M)$ (which is a diffeomorphism onto its image)
2. $\Lambda^*(g'_{ab} \upharpoonright_{\Lambda(M)}) = g_{ab}$
3. $\Lambda(M) \neq M'$.

I will say that spacetime is maximal (or inextendible) iff it has no extension, and φ-maximal (or φ-inextendible) if it has no extension with the property φ. In case of spacetimes I will describe below φ will be a causality condition of global hyperbolicity. Motivation for requiring inextendibility seems to be based on metaphysics (see Earman 1995); nevertheless, inextendibility is commonly taken to be a necessary condition for a spacetime to be physically reasonable.[1]

---

[1] See, however, Manchak (2016b) for a dissenting view.

11.1.3 Levels of Indeterminism in Classical General Relativity

General relativistic spacetimes have many unusual and suprising features. It is appropriate to place indeterminism I am interested in here in a broader context. Depending on how exactly one conceptualizes indeterminism, some general relativistic phenomena count as examples of indeterminism or not. Here, I am interested in situations in which for an initial value problem a solution exists, but uniqueness of a solution fails. But this is by far not the only choice (even if we restrict indeterminism to an existence statement about indeterministic model, which, again, is not a neutral decision). One may distinguish few types of indeterminism in classical GR.

1. acausal solutions, i.e. spacetimes which are not even temporally orientable
2. there is no globally hyperbolic region at all: (a) because closed time-like curves exist, or (b) spacetime satisfies some causality condition which rules out closed timelike curves (strong causality), but is not globally hyperbolic (for example, due to boundary at timelike infinity);
3. globally hyperbolic region is not maximal in the sense that spacetime has been "mutilated" in some way (for example by removing some points from the spacetime manifold);
4. spacetime is singular—it abruptly "comes to an end", either (a) due to blowup of some quantities, or (b) due to some form of curve incompleteness;
5. spacetime does have the maximal globally hyperbolic region, but this region can be further extended in multiple non-equivalent ways

Even though there are many examples of type 1., 2. and 3., they are sometimes argued to be unphysical in classical GR (for example, due to presence of artificial "hole" in spacetime, or on some other grounds). Type 4. spacetimes are rampant in classical GR, as witnessed by the singularity theorems. They may be though of as indeterministic in the sense that solution cannot be continued beyond certain region. Indeterminism is by far not the only possible reading. These spacetimes could be understood as signalling breakdown of the theory (option often assumed by physicists working in quantum gravity), or as entirely new structure predicted by a theory (see Curiel 1998 for this view).

Whether any of these examples could satisfy a Butterfield-like definition depends on the choice of region **S** and the role that the region is expected to play in the analysis. For instance, if **S** is a Cauchy surface for the whole spacetime, only types 3. and 4. remain. And if the role of **S** should be something like: being a candidate for region which determines what happens in its domain of dependence, then 3. seems to be less relevant. But if—as I have argued in Sect. 11.1.1—maximal globally hyperbolic development is a promising choice of **S**, only types 4. and 5. could be of relevance.

Some authors, such as Kutach (2013), restrict their discussion of determinism in GR to singular spacetimes. I find this state of affairs unfortunate, since it ignores fascinating and subtle physical features of various general relativistic spacetimes.

In particular, spacetimes of type 5. have not, I believe, received the attention they deserve—despite being the closest example to the kind of indeterminism which motivates Laplacian concept of determinism in the first place.

There are plenty of ways of mutilating MGHDs and obtain situations in which non-maximal globally hyperbolic region of spacetime can have few extensions: the maximal globally hyperbolic one and some others. Multiple examples of such construction can be found in Manchak (2015) or Earman (1995) chapter 3.8. Such spacetimes are often thought of as artificial, in that some "unphysical" hole has been made in spacetime. It is not easy, however, to find a hole-freeness condition which gives an intuitively correct verdict in such cases (see Krasnikov (2009) and Manchak (2016a) for discussion of few such conditions). But I suppose here that one does not make recourse to any such operations, and allows the globally hyperbolic extension to be continued as far as possible, obtaining MGHD.

Philosophical analysis of GR could be carried at several levels (Malament 2012): one could look for some features at the level of an exact solution, where the right hand side of Einstein's field equation takes some particular form (say, vacuum, electrovacuum, dust, perfect fluid, etc.). Or one could not assume any particular form of the stress-energy-momentum tensor, but assume that it behaves in a certain way by postulating an energy condition. Finally, one could ignore the right hand side entirely, and work with any Lorentzian metric whatsoever.

In what follows I will focus on exact vacuum solutions. Due to the level of technical difficulty, question of existence and uniqueness of extendible MGHD has not been yet investigated thoroughly beyond vacuum solutions, and even in that case the analysis is restricted to certain special (highly symmetric) cases.

11.1.4 Initial Value Problem in GR and Choquet-Bruhat and Geroch Theorem

Under certain conditions initial value problem in classical general relativity has a unique solution, which gives some support to the idea that classical general relativity can be Laplacian deterministic (if one has some reason to ignore spacetimes which are not amenable to initial value treatment due to violations of global hyperbolicity). Choquet-Bruhat and Geroch theorem is a fundamental result expressing this uniqueness of a solution. Informally, one tries to think of a spacelike hypersurface Σ of a 4-manifold (more generally, n-dimensional manifold) $\langle M, g \rangle$ and some data on Σ as uniquely fixing spacetime $\langle M, g \rangle$.

More precisely, initial data consist of an 3-manifold Σ (thought of as spacelike hypersurface of a sought-for M; in full generality, Σ is $n-1$-dimensional, for n being dimension of spacetime) with a Riemannian metric g_0, symmetric covariant 2-tensor k_0 (and, optionally, some scalar functions φ_0, φ_1, etc.). The task is to construct an 4-manifold M with a Lorentz metric g_{ab} (and, optionally, scalar function φ, etc.)

and an embedding $i : \Sigma \rightarrow M$ such that if k is the second fundamental form on $i(\Sigma) \subset M, \ldots$, then $i^*(g) = g_0$, $i^*(k) = k_0$ (and, optionally, $i^*(\varphi) = \varphi_0$, etc.).

For $\langle M, g, k \rangle$ developed from $\langle \Sigma, g_o, k_0 \rangle$ to satisfy Einstein's equations, constraint equations should be satisfied. For a vacuum solution, they take the form:

$$S - k_{0\,ij}k_0^{ij} + (tr\ k_0)^2 = 0$$

$$D^j k_{0\,ji} - D_i(tr\ k_0) = 0$$

where S is the scalar curvature, D is the derivative operator induced on Σ, and indices are raised/lowered by g_0. Then, $\langle \Sigma, g_o, k_0 \rangle$ that satisfies the constraints are called vacuum initial data set.

Choquet-Bruhat & Geroch theorem (Choquet-Bruhat and Geroch 1969), then, states that for an initial (vacuum) data set $\langle \Sigma, g_0, k_0 \rangle$ there exists a unique, up to isometry, maximal globally hyperbolic development (MGHD) $\langle M, g, \Lambda \rangle$ of this data set. "Vacuum" means that the MGHD is a vacuum solution, and the spacetime obtained is a "Cauchy development", that is, $\Lambda(\Sigma)$ is a Cauchy surface in M. The theorem can also be generalized to non-vacuum data sets (see Ringström 2009). This uniqueness result crucially depends on the causality condition (global hyperbolicity) and does not establish uniqueness of the maximal extension of spacetime simpliciter, which will be crucial in the following sections. The point could be stated as follows: φ-inextendible spacetime can be extendible, if its extensions do not satisfy the condition φ.

Final note: this formulation assumes that a spacelike hypersurface has been fixed. But one could just as well fix a null hypersurface and consider initial data living on it. Investigation of this formulation could shed interesting light on the question of determinism of GR, but—since the spacelike (or ADM) formulation is commonly adopted in classical and canonical quantum gravity—I will focus on the spacelike formulation.

I am now in position to inquire whether classical general relativity is deterministic according to Butterfield's definition, after **S** has been decided to be the maximal globally hyperbolic development. In other words, assuming that MGHD exists and is realized in spacetime, can one find any models of the theory which are past (or future, or in both ways) indeterministic, up to certain standard of equivalence of models? Choquet-Bruhat and Geroch theorem does assert uniqueness of MGHD, but remains silent about uniqueness of the spacetime as a whole: from the theorem it does not follow that the non-globally hyperbolic extension of MGHD is unique. If there is a spacetime which has a non-globally hyperbolic extension, it could be unique or non-unique (up to some equivalence relation). So: is there a spacetime A which is the MGHD of some initial data, for which there exist two or more non-globally hyperbolic spacetimes $B_i, i \in I$ into which A can be embedded as a proper subset?

If the answer is "yes", one has a very strong witness of indeterminism. Strength of the witness stems from the fact that even if acausal spacetimes are ignored, and singular behavior is acknowledged as physical feature of the theory, and no artificial manipulations are performed, and the natural course of events is allowed to continue as far as possible, in some cases there is an open possibility of more than one history compatible with given initial data and laws of nature (constraints given by Einstein's field equations).

And the answer, indeed, is "yes, there exist spacetimes such as A". The rest of this paper concerns few consequences of existence of such spacetimes. I discuss the ways in which non-isometric extensions of MGHD arise, to what extent such spacetimes can be considered to be physically reasonable solutions, and implications for the question of Laplacian determinism in classical general relativity.

11.2 Extendible MGHD

We will briefly summarize the current state of the art on the extendible MGHDs. These results are well known in the relevant physical literature (see Chruściel and Isenberg 1993; Ringström 2009), but for our discussion it is useful to have an overview of the situation.

The easiest known example is the Misner spacetime. Take a subregion of two-dimensional[2] Minkowski spacetime (such as a lecture room), close it in the spatial direction (i.e. identify right and left wall of the room), and start contracting it along the closed dimension at some fixed rate β. Take the reference frame (x, t) at rest with respect to the left side of spacetime, and place the clock there. On the left side: $\tau = t$, on the right side $\tau = t/\sqrt{1 - \beta^2}$ (by time dilation). This will lead to creation of closed timelike curves, and chronology horizon (which separates region of spacetime with CTCs from region without CTCs). There are two classes of geodesics: leftward and rightward. Each of these classes can be extended beyond the horizon; but not both of them. In other words: spacetime is geodesically incomplete, but has "less" incompleteness. Misner spacetime is sometimes quoted as an example of spacetime which allows non-isometric extensions, but this is not quite correct: unless some additional requirements on the function which identifies extensions are specified, Misner spacetime has unique extension (a "flip" is needed to map incomplete geodesics in extension I to incomplete geodesics in extension II, and vice versa for complete geodesics).

Taub-NUT spacetime can be seen as a 4-dimensional version of Misner spacetime. Lets begin with $M \approx S^3 \times (t_1, t_2)$, and $g_{ab} = -U^{-1}dt^2 + (t^2 + L^2)(\sigma_x^2 + \sigma_y^2) + 4L^2 U \sigma_z^2$, where $U(t) = \frac{(t_2-t)(t-t_1)}{t^2+L^2}$. Taub spacetime is maximally globally hyperbolic; it could be extended (by adding so-called NUT regions) to the past in two ways and to the future in two ways. In total there are four extensions of the

---

[2]Sticking to the convention followed by most presentations of Misner spacetime in the literature, I discuss two dimensional Misner spacetime. But this spacetime can be defined in higher dimensions as well.

Taub region: $M^{\downarrow+}$, $M^{\downarrow-}$ and $M^{\uparrow+}$, $M^{\uparrow-}$. In every past (future) extensions one of families of geodesics which are incomplete in the Taub region becomes complete, whereas the other family of geodesics remains incomplete (a behaviour analogous to "leftward" and "rightward" geodesics in Misner spacetime). Some of these are isometric: $M^{\downarrow+}$ is isometric with $M^{\downarrow-}$, and $M^{\uparrow+}$ is isometric with $M^{\uparrow-}$. But one can "glue together" pairs of these extensions, one to the past together with one to the future. "Gluing together" means taking a quotient by an appropriate equivalence relation. One obtains four Taub-NUT spacetimes, $M^{++} = M^{\uparrow+} \cup M^{\downarrow+}/ \approx$, $M^{+-} = M^{\uparrow+} \cup M^{\downarrow-}/ \approx$, $M^{-+} = M^{\uparrow-} \cup M^{\downarrow+}/ \approx$, and $M^{--} = M^{\uparrow-} \cup M^{\downarrow-}/ \approx$. Now, how to produce non-isometric extensions of Taub region? It turns out that M^{++} is NOT isometric to M^{+-}, even though M^{++} is isometric to M^{--} and M^{+-} is isometric to M^{-+}. These non-isometric extensions are produced out of isometric ones. But there is no trick involved. Remember that there are two families of null geodesics: in each extension one is complete whereas the other is not. Isometry "exchanges" these two types—this can be done when we have only one extension, but cannot be done (in a smooth way) when we have more of them. So non-isometric extensions of Taub-NUT have to be what I called earlier indeterminism "in both ways".

Misner and Taub-NUT spacetimes have been called counterexamples to almost anything. Are there any other example of maximal globally hyperbolic spacetimes with maximal non-equivalent extensions? Yes. These have been found in (certain) polarized Gowdy spacetimes. These are maximally globally hyperbolic, expanding, not homogenous vacuum spacetimes with two orthogonal Killing vector fields, which (historically) has been useful as simplest toy models of gravitational waves. There are four types of polarized Gowdy spacetimes, classified by topology of the spatial slice: S^3, T^3, $T^2 \times T^1$, and Lens spaces. Depending on the topology of the polarized Gowdy spacetime, it may or may not be extendible. The extendible ones have $M \approx \mathfrak{R}^+ \times T^3$, with a T^2 spatially acting isometry group. Spacetime metric $g_{ab} = e^{-2U(t,\Theta)}[e^{-2A(t,\Theta)}(-dt^2 + d\Theta^2) + t^2 dy^2] + e^{2U(t,\Theta)}dx^2$. These spacetimes are generated by the set of triples $\langle \Pi(\Theta, t), \omega(\Theta, t), \alpha \rangle$, and the behavior near $t = 0$ is governed by $\Pi(\Theta)$. Polarized Gowdy spacetime can be extended iff there is a non-empty interval $I = (\Theta_1, \Theta_2)$ s.t. $\forall \Theta \in I$ $\Pi(\Theta) = 0$ (or $\forall \Theta \in I$ $\Pi(\Theta) = 1$). If spacetime is generated by Π such that $\forall \Theta \in (0, 2\pi)$ $\Pi(\Theta) = 0$ (or $(\forall \Theta \in (0, 2\pi)$ $\Pi(\Theta) = 1$) can be slightly extended ("by ϵ") below $t = 0$. Such a spacetime has two extensions, M_ϵ^+ and M_ϵ^-, which are isometric. But if the spacetime is generated by Π such that Π equals 0 (or 1) in two disjoint intervals, I_1 and I_2, then it has $4 = 2 \times 2$ extensions M_ϵ^{1+}, M_ϵ^{1-}, M_ϵ^{2+}, and M_ϵ^{2-}. Consider the following pasting (by appropriate equivalence relations): $M^{++} = \{M \cup M_\epsilon^{1+} \cup M_\epsilon^{2+}\}/\equiv$, and $\hat{M}^{++}$—the maximal extensions of M^{++}; then $\hat{M}^{++}$ and $\hat{M}^{+-}$ are NOT isometric. Recall now that polarized Gowdy spacetimes are generated by Π. If the number of disjoint intervals on which $\Pi = 0$ or 1 increases, the number of non-isometric extensions spacetimes increases as well. This implies that one can have countably infinite number of non-isometric extensions. All these non-isometric extensions are made to the past (one can think of them as extensions through the Big Bang). Of course, if one decides that polarized Gowdy spacetime should be temporally oriented in the other way, one will have extensions to the future, but the natural interpretation of polarized Gowdy spacetimes sees them as de-idealized cosmological models, whereas change of time orientation would see them as infinite to the past, contracting spacetimes ending with Big Crunch.

Other examples of similar behaviour of MGHD have been found. Ringström (2009) shows that locally rotationally symmetric Bianchi IX initial data have two non-isometric extensions which are C^2 inextendible. He also demonstrates that for other Bianchi types MGHD is inextendible. More recently, Costa et al. (2015) show that for some Einstein-Maxwell systems with positive cosmological constant similar phenomena holds.

11.3 What Do Extendible MGHDs Teach About Classical General Relativity?

If one has extendible MGHD with non-isometric extensions, then one has found indeterminism in the sense of Butterfield's definition—under precisifications concerning choices of region **S** and β I have decided for. That is, there are two models (spacetimes $\langle M, g \rangle$ and $\langle M', g' \rangle$) which contain regions of kind **S** (maximal globally hyperbolic developments) such that there exists an isometry between the MGHDs, but there is no isomorphism β (isometry preserving temporal orientation) between $\langle M, g \rangle$ and $\langle M', g' \rangle$. Therefore classical general relativity is not MGHD-deterministic. This is interesting on its own, since it shows that even if spacetime is globally hyperbolic, there could be physical phenomena (formation of Cauchy horizons) which allow for indeterminism in the sense of the existence of solutions and the failure of uniqueness. But a foundational significance of extendible MGHDs is not limited to that. In what follows I discuss common features of these spacetimes, pressure they put on the simplistic understanding of physical equivalence in GR, whether they are physically reasonable or not, their relation to questions concerning time machines, and relations between cosmic censorship and determinism.

11.3.1 Common Features

Known cases of non-unique extensions share few common features. These spacetimes

1. are exact solutions with non-trivial symmetries,
2. are spatially compact (i.e. topology of the spacelike section is compact),
3. are vacuum solutions (i.e. right-hand side of the Einstein's field equations vanishes),

4. do not satisfy strong causality in the non-isometric regions (a topic I return to in Sect. 11.3.5),
5. are singular
6. and these which have non-unique up to time preserving isometry extensions have disconnected Cauchy horizons (i.e. the Cauchy horizon has more than one connected component: one to the past and one to the future in Taub-NUT case, multiple components separated by curvature singularities in Gowdy case)

There is no proof that any of these needs to hold. It is just a fact that all MGHDs with non-unique extensions found so far share these properties.

Costa et al. (2015) found electrovacuum solutions with positive cosmological constant which admit non-unique extensions. This shows that feature 3. is not necessary for the MGHD to have non-unique extensions. Similarly, one may expect that features 1. and 2. could be relaxed – these features, then, arise merely due to our lack of control over the set of relativistic spacetimes (in contrast with being a property that MGDHs with non-unique extensions need to satisfy).

Note, moreover, that the fourth feature holds only if some form of hole freeness condition is assumed – otherwise, one can always remove closed regions from the extension and use conformal transformations or pass to universal covering space to ensure inextendibility, while simultaneously removing closed time-like curves from the extension (graphically speaking, by cutting them into half in the process of removal of the closed regions). I expand on this theme in Sect. 11.3.3, linking this question to some difficulties encountered in investigations concerning the existence of time machines in classical general relativity.

Fifth feature raises a question about relation between spacetime singularities and indeterminism. If, as I assume here, one conceptualizes indeterminism as situation in which solutions exist but their uniqueness fails, then singular spacetime does not have to demonstrate indeterminism of the theory. In case of "singular" spacetimes, there is **no** evolution of system for some parameter; in case of extendible MGHDs, there is **more than one** evolution of the system compatible with the given initial data. But all cases of extendible MGHDs are singular in the sense of being timelike geodesically incomplete (and it is difficult to imagine that this particular feature could be relaxed, at least in case of vacuum solutions): any of the inextendible non-unique extensions of Misner, Taub-NUT, polarized Gowdy or Bianchi IX spacetime is timelike geodesically incomplete (even if one allows non-Hausdorff extensions; see Hawking and Ellis (1973, p. 174) for the argument in Misner case). But they do not have to be singular in the sense of curvature blow-up: Misner and Taub-NUT spacetimes are blow-up free.

A topologically disconnected Cauchy horizon is present in all known cases of extensions which are non-unique up to isometry which preserves temporal orientation. Going back to the classification of indeterministic solutions with respect to temporal orientation, one could conjecture that non-uniqueness of a solution requires either that the solution is "in both ways" indeterministic, or that it needs to have curvature singularity in some (but not all) regions close to the Cauchy horizon, effectively splitting it into at least two components. And one could also try to play some forms of singular behaviour against some other forms: for example, if it could be demonstrated that Cauchy horizons can be replaced with blow-up singularities (a theme explored by Misner and Taub in the context of classical instability in Misner and Taub (1969), and, in the context of semi-classical gravity, by Hiscock and Konkowski (1982) for Taub-NUT horizons and Thorne (1993) for Misner spacetime), then one type of singularity (more bening, so to speak) could be used to protect us from more malign phenomena.

11.3.2   Physical Equivalence

One often hears that diffeomorphic spacetimes are equivalent, or other statement to that effect (these are made in the context of debates concerning the Hole Argument). An interesting concern raised by existence of extendible MGHDs is that mere existence of an isometry between spacetimes may not be sufficient condition for physical equivalence. Consider following conditions:

1. Isometry preserves time orientation
2. Isometry belongs to the connected component of the identity in the diffeomorphism group (with the intuition being that only isometries which can be obtained by continuous deformation from an identity represent physically equivalent situations)
3. Isometry preserves location of a Cauchy surface[3]

These conditions are sensitive to whether $\langle M, g \rangle$ has non-isometric extensions: a spacetime can have unique extension if one uses weaker (i.e. broader) condition, but non-unique if one uses stronger (more strict) condition. For instance (Chruściel and Isenberg 1993) if one considers 2-dimensional Misner spacetime and demands that isometry is in the connected component of identity, then maximal extension is non-unique. Thus, whether a given spacetime counts as an example of indeterminism or not can be changed by demanding more fine-grained equivalence condition. Similarly, extensions to the future (or to the past) of the Taub region are isometric by a time orientation preserving isometry, but not by an isometry belonging to the identity component of the diffeomorphism group. And in all these cases extensions are not Cauchy equivalent. But is there a single correct condition for physical equivalence which should be used in the context of initial value problem? Equivalently, do these more strict conditions capture some physical structure which should be preserved by mappings which identify physically equivalent spacetimes? In particular, are there theoretical contexts in which these strict conditions are appropriate?

---

[3]More precisely: two maximal extensions $\langle M_1, g_1, \Lambda_1 \rangle$ and $\langle M_2, g_2, \Lambda_2 \rangle$ of $\langle M, g \rangle$ are Cauchy equivalent wrt Σ, i iff $\psi \circ \Lambda_1 \circ i = \Lambda_2 \circ i$

Note also that choice of a differentiability class does not seem to influence whether spacetime has non-isometric extensions. In case of Taub-NUT and polarised Gowdy, extensions are analytic; in case of Bianchi IX extensions are C^2. This makes the extensions slightly different from Norton's dome, where differentiability conditions (singularity at the summit) play crucial role.

11.3.3 Distinguishing Physically Reasonable Spacetimes from Physically Unreasonable Ones

In the context of indeterminism of Newtonian mechanics (as demonstrated by the Norton's dome) Malament (2008) suggested that interesting question raised by Norton's dome is not a yes-or-no question whether Newtonian mechanics is indeterministic; rather, the interesting question is to whether and in what sense examples discussed by Norton can be thought of as examples of Newtonian systems. Similar question could be asked about extendible MGHDs: to what extent are they examples of physically reasonable general relativistic spacetimes?

This leads to a more general point: analysis of determinism by asking about existence and properties of extendible spacetimes is closely related to questions concerning physical possibilities in classical general relativity. Consider some Problematic Feature φ of the theory; in our case the feature is indeterminism, but in other cases it may be something else, like singular structure, observational indistinguishability, or possibility of time travel. Questions concerning Problematic Feature φ can be thought of as investigated in two stages. In the first stage, one questions whether φ is possible by looking at all spacetimes $\langle M, g \rangle$. It turns out that it is, in general, very easy to find some examples of spacetimes with property φ, for most properties which are of some philosophical interest. Thus, in the second stage, one considers only those spacetimes $\langle M, g_{ab} \rangle$ which satisfy some Nicety Condition, which distinguishes physically reasonable spacetimes from physically unreasonable ones. Whether it is still easy to find spacetime with property φ depends on the form of the additional condition.

Since there are many senses in which a spacetime can be unphysical, there is no single Nicety Condition appropriate for all contexts. For example, in cosmological context the Nice Feature often boils down to the demand of isotropy and homogeneity, whereas for some other contexts, such as possible distributions of matter, this demand would be much too strong.

Various conditions have been used as a Nicety Condition: being a solution of Einstein's equations is an obvious one; closely related are energy conditions; less obvious ones are various inextendibility and hole freeness-type conditions; and stability conditions (since stability is related to an important issue of cosmic censorship, I discuss it separately in Sect. 11.3.4). One thing is clear: known examples of extendible MGHDs are not empirically adequate as a representation of the large scale structure of the universe we live in. In this sense classical general relativity is indeterministic, even though the world as described by the theory could be deterministic.

Einstein's equations are satisfied in the case of extensions I discussed earlier. But there is a sense in which Einstein's equations are a vacuous constraint anyway: whatever the metric g_{ab} is, it can always be used to define some stress-energy-momentum tensor on the right hand side. Very often the right hand side obtained in this way will be "unphysical", for example because energy density will be negative. Energy conditions are then postulated to distinguish between stress-energy-momentum which are reasonable and those which are not. But (see Curiel 2014) the status of energy conditions seems to be highly problematic, as it is very easy to produce examples of classical and quantum scalar fields which violate any known energy condition. In addition, when a distinction between reasonable and unreasonable spacetime is drawn using an energy condition, even though it is assumed the former ones satisfy some energy condition, an explicit choice of a particular condition is very rarely made.

Another group of conditions are those which require that spacetime is "as large as it could be"—geodesic completeness, hole freeness, inextendibility, and similar. Should extendible MGHD be counted as physically possible according to these conditions? All extensions of extendible MGHDs have incomplete geodesics (under the assumption that spacetime is a smooth Hausdorff manifold; see Müller and Placek (2015) for a dissenting view). But most solutions of GR have incomplete geodesics, due to singularity theorems. Whether hole freeness is satisfied or not depends on the details of the definition. But since most available definitions are variations on the theme of the domain of dependence being as large as possible, and since one always has MGHD realized in the case of spacetimes I described above, these hole freeness conditions cannot be used to dismiss indeterminism present in case of extendible MGHDs as physically unreasonable. Finally, inextendiblity condition actually motivates going beyond MGHD (since in these cases MGHD is as large as it could be, but the spacetime as a whole is not as large as it could be). Insofar as inextendibility is motivated by something like a principle of sufficient reason, this principle forces us to leave the calm waters of MGHD and accept wild indeterminism of non-unique regions.

Of course, any non-unique, inextendible non-globally hyperbolic extensions of maximally globally hyperbolic spacetime violates some causality condition (for instance, if they are non-unique, they cannot be globally hyperbolic). But dismissing these extensions on the grounds of violating causality condition *by fiat* seems question-begging: by doing so, one would effectively assume that physically reasonable spacetimes need to be deterministic.

And if one would hold the view[4] that a spacetime is physically reasonable iff it could be used to model actual physical phenomena, then—since Gowdy spacetimes have been developed as toy models for studying gravitational waves—some extendible MGHDs would count as physically reasonable according to such a criteria.

---

[4]I am not aware of anyone actually subscribing to this in writing.

To wrap up: from most commonly used conditions for distinguishing physically reasonable spacetimes from unreasonable ones, all those which are not question-begging are satisfied by non-isometric extensions of MGHDs. Thus, there is no easy dismissal of the form of indeterminism brought by extendible MGHDs as physically unrealistic.

11.3.4 Rarity, Stability, Cosmic Censorship and Determinism of General Relativity

Cosmic censorship (Penrose 1979) is the statement that physically reasonable spacetime is globally hyperbolic. And if, as Penrose suggest, spacetime is physically unreasonable iff it is unstable against some time of perturbations, there are prospects for a rigorous proof of the cosmic censorship. More recent statement takes the following form: generic initial data have inextendible MGHDs. This version is more precise, but still not precise enough, since there are multiple ways of specifying what is generic. Ringström (2010) attributes this statement of cosmic censorship to Chrusciel, hence I will call it Chrusciel's Strong Cosmic Censorship, CSCC (even though similar statements appear in literature before, for example in Moncrief 1981).

In a similar vein, Hawking (1971) argued that only properties stable against some type of perturbations are physically relevant. But being generic depends on topology one chooses, and Fletcher (2015) shows that there are different choices of topologies available, each coming with its own shortcomings. So if one takes the property of "having MGHD extendible in non-unique way", identifies being generic with being stable against perturbations, and finds a physical justification for the choice of the particular topology (all of which are non-trivial conceptual choices), and in that topology statement expressing CSCC holds (see Ringström (2010) for a summary of known results), then spacetimes I have been discussing are not physically reasonable witnesses of indeterminism. Note that known theorems do not assert a typicality result in a given measure, but mere a typicality in given topology.

There seem to be few types of rarity present in extendible MGHDs. First, these spacetimes have non-trivial global symmetries, and it is widely accepted that highly symmetric spacetimes are non-generic. Second, in case of Misner spacetime a peculiar geometric setup (with contraction along the spacelike axis) is present, which presumably could be used to argue that this spacetime is non-generic. Third, initial data whose MGHDs are extendible polarized Gowdy spacetimes need to contain analytic functions Π which take the value 0 (or 1) on two or more disjoint intervals, which are topologically atypical. Under these particular understandings of "generic", CSCC holds for known extendible MGHDs.

Lets now go back to the notion of Laplacian determinism introduced at the beginning. This notion does not care about size or relative placement of indeterministic solutions in the space of all solutions. Indeterminism is merely an existence statement: if there are at least two solutions which have isometric region **S** but are not isometric overall, the theory is indeterministic. CSCC, on the other hand, does not care about existence statements like this, unless they are being followed by a typicality statement. This leads to a disconnect: even though the examples I discussed show that GR is indeterministic in the sense of Butterfield's definition, they are compatible with CSCC. Since Laplacian determinism fails and CSCC holds, CSCC cannot express Laplacian determinism.

Is there, then, some other notion of determinism which CSCC expressess? One could try to understand it as expressing a form of well-posedness in the Hadamard sense. Well-posedness in the Hadamard requires that solutions exist, are unique, and moreover that solutions continuously depend on the initial data, meaning that small changes to the initial data result in small overall changes in the solution. Note, however, that here uniqueness fails in "rare" cases, and so well-posedness expressed by CSCC must be somewhat non-standard. It remains to be seen whether there is a robust, conceptually well-motivated notion of determinism which would find formal expression in such a non-standard notion of well-posedness. Certainly notion of determinism discussed in the philosophy of science literature has nothing of it.

So far, I have focused on vacuum solutions with cosmological constant Λ set to zero. What happens if Λ is non-zero? Earman (1995)(p. 82) noted that "it remains to be seen whether or not, on balance, the cosmological constant helps or hinders the quest for cosmic censorship". I will that the tentative answer is: it hinders the quest. In an interesting recent development, Costa et al. (2015) argue that positive cosmological constant makes things worse, in the sense that they are able to find spherically symmetric characteristic initial data for the Einstein-Maxwell-scalar field, with positive Λ, such that there are non-isometric and generic C^1 extensions through the horizon of the Reissner-Nordström black hole. Since negative cosmological constant often makes CSCC inapplicable (due to non-existence of MGHD in spacetimes with asymptotically anti-de Sitter behaviour at infinity), on balance it seems that cosmological constant leads to more indeterminism (although, in the negative case, it is not the Laplacian determinism in the sense I explicated in Sect. 11.1.1.

11.3.5 Connections Between Some Forms of Indeterminism and Existence of Time Machines

At this point careful reader may ask the following: aren't all known extendible spacetimes instances of Thornian time machines, that is, spacetimes which do not contain closed time-like curves, but in which the arrangement of matter and geometric features brings about closed time-like curves?[5,6] And if so, is it not

---

[5]Similar questions could be asked about Malament-Hogarth spacetimes.

[6]For difference between Thornian and Wellsian time machines, see Earman et al. (2016).

the case that question of determinism is subsumed by question whether time machines are genuine physical possibilities?

There are various ways of making the notion of a time machine precise (see Earman et al. (2009) for discussion of these issues). A precisification suggested by Earman, Smenk and Wüthrich (and often called ESW time machines in subsequented literature, cf. Manchak (2009) or Manchak (2014)) demands, intuitively, that region in which time machine operates is free from closed time-like curves (and MGHD, of course, is CTC-free), and every extension to the future of that region contains closed time-like curves.

Extensions of Misner and polarized Gowdy spacetime with reversed temporal orientation are extensions of the sort one is interested when investigating existence of ESW time machines. But extensions of Taub-NUT spacetime (which are "in both ways") and extensions of polarized Gowdy spacetime with natural temporal orientation (which are made to the past) resemble a reversed time machine: there are CTCs before the Big Bang, but once an observers passes through the horizon to MGHD, there is no way back to the region which allows time travel.

Moreover, since one cannot presently exclude existence of extendible MGHD such that all of the non-isometric extensions satisfy strong causality condition (which prevents CTCs), it does not follow that every spacetime admitting non-isometric extensions to the future is an example of spacetime allowing for operation of a time machine. Additionally, one can perform various mathematical operations in the non-globally hyperbolic region: for instance, remove various subsets of the manifold in such a way that there will be no CTCs in the resulting spacetime. If such extensions are allowed, the number of non-isometric extensions available increases dramatically. Resulting spacetime will have unique MGHD with various extensions: some with CTCs, some without. Most of such extensions would, intuitively speaking, have artificial "holes"; but, again, it is rather difficult to find a formal hole freeness condition which gives correct verdict in these cases.

11.3.6 Open Question

All of this leads to an interesting open question. Is there any spacetime which is maximal globally hyperbolic, satisfies some energy condition and admits at least two non-isometric extensions which are (a) hole free (in a yet to be made precise sense, if that is possible), (b) not Malament-Hogarth, (c) satisfy at least strong causality, (d) extensions are to the future only, (e) Cauchy horizon has single connected component, (f) satisfies some energy condition? Is there a robust or stable (in some well-defined sense) class of such spacetimes?

Of course, even if such a robust class of spacetimes were found, it would not mean that our world is indeterministic. Classical general relativity is often conceptualized as having "sectors" (that is, space of solutions is divided into various subsets, where spacetimes lying in a given subset share some relevant features – for example, symmetry group, form of the stress-energy-momentum tensor, or a value of the cosmological constant), and maybe the most useful representations of our world would be found in the deterministic sector. So finding a class of extendible spacetimes satisfying above constraints which could serve as a useful model for the universe we actually seem to live in would be the ultimate triumph of indeterminism in classical GR.

11.4   Summary: Laplacian Indeterminism in Classical General Relativity

Examples of extendible MGHDs can be interpreted in two ways:

(a) classical GR is indeterministic
(b) classical GR is deterministic: extendible MGHD are not physically relevant

I have argued that option (b) is untenable: there is no question-begging condition for being physically reasonable spacetime which is violated by extendible MGHDs; some of these spacetimes (polarized Gowdy solutions) have been used as useful toy models, and we have no reason to expect that phenomena of extendible MGHD with non-unique extension is limited to toy models only; and foundational significance of stability condition (not to mention difficulties with finding a formal expression of such a condition) is murky at best.

Since option (b) seems untenable, one is left with indeterminism of the theory (again, in the sense of having some indeterministic models). This assessment agrees with sentiments (sometimes) expressed both by philosophers – Belot (2011):

> Instances in which globally hyperbolic solutions admit non-isometric extensions are instances of genuine indeterminism, not gauge equivalence

and physicists – Ringström (2009):

> The fact that there are inequivalent maximal extensions [of maximal globally hyperbolic developments] means that the initial data do not uniquely determine a maximal development. In this sense, the general theory of relativity is not deterministic.

or Ringström (2010), where he comments that extendible MGHDs:

> demonstrate that Einstein's general theory of relativity is not deterministic; given initial data, there is not necessarily a unique corresponding universe

(and goes on to discuss the consolation brought by rarity of such spacetimes, i.e. Chrusciel's version of the cosmic censorship).

I have argued that it is not straightforward to understand the strong cosmic censorship in Chrusciel's sense as expressing determinism. But if it does not express determinism, what does it say?

I suggest that it merely expresses the hypothesis that GR is not radically indeterministic: there may be solutions which are indeterministic, but there are few and far between. Similar theme can be found in Geroch (1970) metaphor of black and white patches on the paper (where paper represents the set of all spacetimes, black patches are singular spacetimes and white patches are spacetimes which are no singular; the question, then, is whether from a large distance paper looks rather white, black, or kind of greyish; and Penrose and Hawking's singularity theorems are taken to imply that it is at least dark greyish). I propose that existence of extendible MGHDs be interpreted in the very same way; in particular, cosmic censorship in Chrusciel's formulation does not express determinism of classical GR, but rather a well-formed mathematical hypothesis that classical general relativity (under auxiliary assumptions, such as lack of holes and restriction to spacetimes which are amenable to initial value formulation at all) is not radically indeterministic. Using Geroch's metaphor concerning patches of paper, cosmic censorship hypothesis states that extendible MGHDs are rare enough that the set of solutions is light grey or white when looked at from the distance. Lack of radical indeterminism may be the next best thing, but is not the same as determinism *simpliciter*.

One could object that study of philosophical consequences of extendible MGHDs is a waste of time: after all, even if they are indeterministic models of the theory, they are far from being a most fitting to the data representation of the observable universe. So why bother? I would say that there are at least two reasons. First, that we are in philosophically interesting situation in which the observable universe is best represented as a deterministic model of an otherwise indeterministic theory. Second, that we can and do learn a lot about the theory by studying its various solutions, and extendible MGHDs are useful for that purpose; in particular, they demonstrate subtle ways in which various forms of indeterminism and singular behaviour of spacetime can mingle with each other.

Acknowledgements The author acknowledges the funding of the research grant 'Mistrz 2011' of the Foundation for Polish Science, contract nr 5/2011 for financial support. The work on this paper was partically carried out as a part of the "Probability, Causality and Determinism" Bilateral Mobility Grant of the Hungarian and Polish Academies of Sciences, NM-104/2014. I would also like to thank Dennis Dieks, Erik Curiel, Sam Fletcher, F. A. Muller, Christian Wüthrich, and the audience at BSPS 2014 in Cambridge for helpful feedback. Most of this material has been developed through numerous conversations with Tomasz Placek, to whom I am very grateful.

References

Belot, G. 2011. Background-independence. *General Relativity and Gravitation* 43: 2865–2884.
Butterfield, J. 1989. The hole truth. *British Journal for the Philosophy of Science* 40(1): 1–28.
Choquet-Bruhat, Y., and R. Geroch. 1969. Global aspects of the cauchy problem in general relativity. *Communications in Mathematical Physics* 14(4): 329–335.
Chruściel, P.T., and J. Isenberg. 1993. Nonisometric vacuum extensions of vacuum maximal globally hyperbolic spacetimes. *Physical Review D* 48(4): 1616.
Costa, J.L., P.M. Girão, J. Natário, and J.D. Silva. 2015. On the global uniqueness for the Einstein–Maxwell-scalar field system with a cosmological constant. *Communications in Mathematical Physics* 339(3): 903–947.
Curiel, E. 1998. The analysis of singular spacetimes. *Philosophy of Science* 66: S119–S145.
Curiel, E. 2014. A primer on energy conditions. In *Towards a theory of spacetime theories*, 43–104. Springer. arXiv preprint:1405.0403.

Earman, J. 1986. *A primer on determinism*, vol. 37. Dordrecht: D. Reidel.

Earman, J. 1995. *Bangs, crunches, whimpers, and shrieks: Singularities and acausalities in relativistic spacetimes*. New York: Oxford University Press.

Earman, J. 2007. Aspects of determinism in modern physics. *The Philosophy of Physics* 2: 1369–1434.

Earman, J., C. Smeenk, and C. Wüthrich 2009. Do the laws of physics forbid the operation of time machines? *Synthese* 169(1): 91–124.

Earman, J., C. Wüthrich, and J. Manchak. 2016. Time machines. In *The Stanford encyclopedia of philosophy* (Winter 2016 ed.), ed. E.N. Zalta. Stanford: Metaphysics Research Lab, Stanford University.

Fletcher, S.C. 2015. Similarity, topology, and physical significance in relativity theory. *The British Journal for the Philosophy of Science* 67(2): 365–389.

Geroch, R. 1970. Singularities. In *Relativity*, 259–291. New York: Plenum.

Hawking, S., and G. Ellis. 1973. *The large scale structure of space-time*. Cambridge monographs on mathematical physics. Cambridge: Cambridge University Press.

Hawking, S.W. 1971. Stable and generic properties in general relativity. *General Relativity and Gravitation* 1(4): 393–400.

Hiscock, W.A., and D.A. Konkowski. 1982. Quantum vacuum energy in taub-nut (Newman-Unti-Tamburino)-type cosmologies. *Physical Review D* 26: 1225–1230.

Krasnikov, S. 2009. Even the Minkowski space is holed. *Physical Review D* 79(12): 124041.

Kutach, D. 2013. *Causation and its basis in fundamental physics*. Oxford: Oxford University Press.

Malament, D. 2012. *Topics in the foundations of general relativity and Newtonian gravitation theory*. Chicago lectures in physics. Chicago: University of Chicago Press.

Malament, D.B. 2008. Norton's slippery slope. *Philosophy of Science* 75(5): 799–816.

Manchak, J.B. 2009. On the existence of time machines in general relativity. *Philosophy of Science* 76(5): 1020–1026.

Manchak, J.B. 2014. Time (hole?) machines. *Studies in History and Philosophy of Science Part B: Studies in History and Philosophy of Modern Physics* 48: 124–127.

Manchak, J.B. 2015. On Gödel and the ideality of time. *Philosophy of Science* 83(5): 1050–1058. Chicago: University of Chicago Press.

Manchak, J.B. 2016a. Epistemic "holes" in space-time. *Philosophy of Science* 83(2): 265–276.

Manchak, J.B. 2016b. Is the universe as large as it can be? *Erkenntnis* 81(6): 1341–1344. Springer.

Misner, C., and A. Taub. 1969. A singularity-free empty universe. *Soviet Physics – JETP* 28: 122.

Moncrief, V. 1981. Infinite-dimensional family of vacuum cosmological models with taub-nut (Newman-Unti-Tamburino)-type extensions. *Physical Review D* 23(2): 312.

Müller, T., and T. Placek. 2015. Defining determinism. *The British Journal for the Philosophy of Science*: axv049. The British Society for the Philosophy of Science.

Penrose, R. 1979. Singularities and time-asymmetry. In *General relativity: An Einstein centenary survey*, 581–638.

Ringström, H. 2009. *The Cauchy problem in general relativity*. Zürich: European Mathematical Society.

Ringström, H. 2010. Cosmic censorship for Gowdy spacetimes. *Living Reviews in Relativity* 13(1): 2.

Thorne, K.S. 1993. Misner space as a prototype for almost any pathology. In *Directions in general relativity: Papers in honor of Charles Misner*, vol. 1. Cambridge University Press.

Werndl, C. 2016. Determinism and indeterminism. In *The Oxford handbook of philosophy of science*, ed. P. Humphreys. Oxford University Press.

Wüthrich, C. 2011. Can the world be shown to be indeterministic after all? In: *Probabilities in physics*, 365–390. Oxford: Oxford University Press.

Chapter 12
How Do Macrostates Come About?

Márton Gömöri, Balázs Gyenis, and Gábor Hofer-Szabó

Abstract This paper is a further consideration of Hemmo and Shenker's ideas about the proper conceptual characterization of macrostates in statistical mechanics. We provide two formulations of how macrostates come about as elements of certain partitions of the system's phase space imposed on by the interaction between the system and an observer, and we show that these two formulations are mathematically equivalent. We also reflect on conceptual issues regarding the relationship of macrostates to distinguishability, thermodynamic regularity, observer dependence, and the general phenomenon of measurement.

Keywords Macrostates • Distinguishability • Thermodynamic regularity

12.1 Introduction

Macrostates are distinguished subsets of a system's phase space. They play an essential role in statistical mechanics since they are identified—at least in the Boltzmann program—with thermodynamic states and thus provide the basis for the statistical mechanical explanation for thermodynamic phenomena. It is a crucial task for statistical mechanics to give a physical explanation for why certain sets of microstates become distinguished, or in short, how macrostates come about.

As an answer to this question, in their illuminating book *The Road to Maxwell's Demon, Conceptual Foundations of Statistical Mechanics* Meir Hemmo and Orly Shenker discern two characteristic features of macrostates:

1. Microstates in a macrostate are indistinguishable, while macrostates are distinguishable to a human observer.

   > "One kind of physical property according to which sets of microstates can be defined is distinguishability by a given observer: in general, observers are unable to distinguish

M. Gömöri (✉) • B. Gyenis • G. Hofer-Szabó
Institute of Philosophy, Research Centre for the Humanities, Hungarian Academy of Sciences,
Budapest, Hungary
e-mail: gomori.marton@btk.mta.hu; gyenis.balazs@btk.mta.hu; szabo.gabor@btk.mta.hu

© Springer International Publishing AG 2017
G. Hofer-Szabó, L. Wroński (eds.), *Making it Formally Explicit*, European Studies in Philosophy of Science 6, DOI 10.1007/978-3-319-55486-0_12

213

between individual microstates, but can distinguish between certain sets of microstates; and each distinguishable set of indistinguishable microstates forms a macrostate." (Hemmo and Shenker 2012, p. 95)

2. Macrostates feature in thermodynamic regularities (such as the ideal gas law).

"Another kind of physical property of interest, shared by microstates, is the one that gives rise to the thermodynamic regularities. Certain sets of microstates exhibit this particular kind of regularity: all the microstates in these sets appear to satisfy the same laws, described by the theory of thermodynamics." (Hemmo and Shenker 2012, p. 95)

These two characterizations of the macrostates are conceptually distinct: the first characterization refers to the system's relation to an external observer while the second one refers only to features inherent to the system. Still, there is a significant coincidence between these two features of the macrostates:

"It is a contingent fact about the structure of human beings as observers, that there is a useful degree of overlap between the sets that satisfy the regularities and the sets that correspond to our observation capabilities." (Hemmo and Shenker 2012, p. 96)

Hemmo and Shenker keep emphasizing that despite being observer-relative macrostates are objective. Their book provides an insightful analysis of what distinguishability means in terms of the physical interaction of the observer and the target system. The key idea of their analysis is that the interaction of the system and the observer brings about a one-to-one correlation between certain sets of microstates in the system's phase space and the observer's phase space. It is these sets of the system's phase space that they call the *macrostates* of the system relative to the observer.

The main aim of the present paper is to provide a precise mathematical description of how macrostates come about via the correlation of states of an observer and a system. We hope that this analysis will contribute to making Hemmo and Shenker's illuminating ideas a bit sharper.

The paper is structured as follows. In Sect. 12.2 we give an intuitive introduction to the idea behind the mechanism of the coming about of macrostates. In Sect. 12.3 we provide a mathematically rigorous analysis of the same mechanism together with a mathematically equivalent characterization of macrostates. In Sect. 12.4 we provide a formal description of the observer-relativeness of macrostates. Readers not interested in technicalities can go directly to Sect. 12.5 where the results of the previous two sections will be summarized in an informal way. In Sect. 12.5 we reflect on conceptual issues regarding the relationship of macrostates characterized in the above way to distinguishability, thermodynamic regularity and the general phenomenon of measurement. We conclude in Sect. 12.6.

12.2   Macrostates: Triggering Intuition

Let O be an observer and S a physical system. Let X_O and X_S denote the *phase space* of O and S, respectively. Generally we will not assume that X_O and X_S have any mathematical structure. They will simply be sets if not explicitly stated otherwise.

Let $O + S$ denote the *joint system* of the observer and the system, and let X_{O+S} denote the phase space of the joint system. We stipulate that a *microstate* $x \in X_{O+S}$ of the joint system is given by a pair (x_O, x_S) where $x_O \in X_O$ is a *microstate* of the observer and $x_S \in X_S$ is a *microstate* of the system; that is $X_{O+S} = X_O \times X_S$ is a Cartesian product. Sometimes we will refer to x_O and x_S as the *projection* of x onto X_O and X_S, respectively. Similarly, for any *microregion* $B \subseteq X_{O+S}$ let B_O and B_S denote the projection of B onto X_O and X_S, respectively.

The possible microstates of the joint system $O + S$ are often confined to a subregion of X_{O+S} due to various physical conditions. Denote this *accessible region* of microstates of the joint system by A and its projections onto X_O and X_S by A_O and A_S, respectively. The exact "shape" of the accessible region A depends on the nature of the physical interaction between the observer and the system and, as we will shortly see, it is playing a crucial role in the coming about of macrostates.

How do macrostates come about?

The central idea, due to Hemmo and Shenker (2012, Ch. 5), is that (i) macrostates of a system S are relative to an observer O; (ii) they emerge from the many-to-many type correlation between the microstates of O and S; and, most importantly, (iii) this correlation is established by the accessible region A of the joint system. In the next two sections we will give a mathematically precise formulation of these ideas; here we just trigger intuition.

Consider Fig. 12.1, borrowed from the book of Hemmo and Shenker.

Here the microstates of O are depicted along the vertical axis of the diagram and the microstates of S along the horizontal axis. The accessible region A of the joint system is a slant line. Note that the axes and the straight line are just for illustration since the phase spaces have no linear, metric or topological structure.

Now, the accessible region A of the joint system is such that it establishes a one-to-one correlation between the microstates of O and S. In other words, the outer physical conditions and the nature of the interaction between the observer and the system are such that if the observer O is in a certain microstate, the system S is forced to be in one particular microstate.

Consider now another situation where the accessible region A of the joint system consists of two disconnected regions, B_1 and B_2. (See Fig. 12.2.) What kind of constraints arise due to A between the microstates of O and S? Obviously, if the

Fig. 12.1 One-to-one correlation between the microstates of O and S in a connected accessible region A

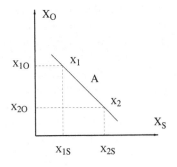

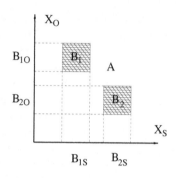

Fig. 12.2 Many-to-many correlation between the microstates of O and S in a disconnected accessible region A

microstate of O is in B_{1O}, then the microstate of S cannot be in B_{2S}. Thus, the microstates of the two subsystems cluster into two groups: microstates of the system in B_{1S} correlate with microstates of the observer in B_{1O}, and microstates in B_{2S} correlate with microstates in B_{2O}.

Observe that the difference between the two cases lies completely in the difference between the two accessible regions. In the first case the accessible region maps the microstates of the observer to the microstates of system in a one-to-one way, whereas in the second case the accessible region maps only one particular partition of the phase space of the observer, namely $\{B_{1O}, B_{2O}\}$ to one particular partition of the phase space of the system, namely $\{B_{1S}, B_{2S}\}$. Stating it differently, in the first case the accessible region maps *any* partition of the phase space of the observer to the corresponding partition of the phase space of the system; whereas in the second case the accessible region maps only *specific* partitions of the phase space of the observer to *specific* partitions of the phase space of the system in a one-to-one manner.

And this is the point where macrostates enter the scene. *Macrostates* are elements of specific partitions of X_S, namely of those partitions which are mapped by the accessible region of the joint system $O + S$ into partitions of X_O in a one-to-one way.

What is the idea behind this definition of macrostates?

Suppose you measure the temperature of a container of gas. Measuring temperature means that one introduces an interaction between a measuring device, the thermometer, and the system. The nature of this interaction fixes which joint microstates the joint system can have. The different mercury levels partition the phase space of the thermometer regarded as a system of particles. But in order for a thermometer to count as a *measuring device* and for the mercury levels to count as *measurement outcomes* it is also needed that the different levels of mercury tell something about the state of the measured system. A given mercury level is compatible with many microstates of the thermometer; and it will not specify the microstate of the system either. However, what is required in the above definition is that the level of mercury should indicate at least in which *set* of microstates the system is. To do this, the sets pertaining to different mercury levels cannot overlap—otherwise the thermometer would not serve as a good measuring device; and they

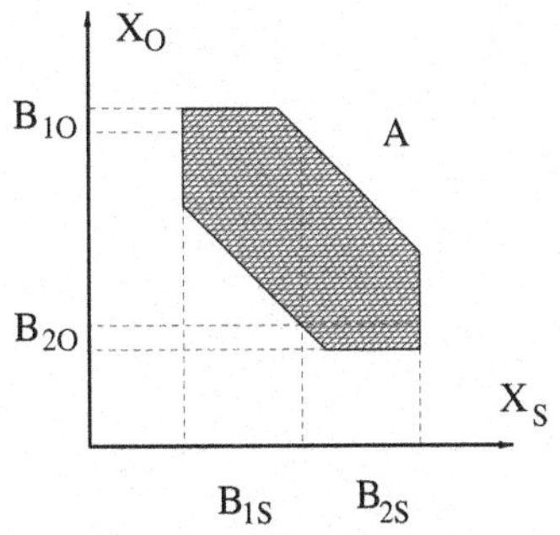

Fig. 12.3 Many-to-many
correlation with disjoint sets
on the phase space of O

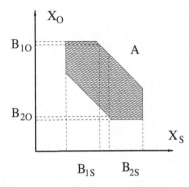

Fig. 12.4 Many-to-many
correlation with disjoint sets
on the phase spaces of
O and S

together should cover the phase space of the system—otherwise some microstates
of the system could not be indicated.

One could, however, also take a more liberal stance towards macrostates.
Suppose that the phase space of the observer cannot be partitioned entirely but there
is a *set of disjoint subsets* such that their image *via* the the accessible region in
the above sense forms a partition of the phase space of the system. (See Fig. 12.3.)
The difference between this case and the previous one is that here the sets of those
microstates of the observer which stand in a one-to-one relation with a certain
partition of the system's phase space do not form a partition. Some microstates of
the observer do not indicate any macrostate of the system.

An even more liberal definition of the macrostates would be to relax the demand
of being a partition both for the phase space of the observer and also for the system
and only to demand that disjoint subsets of O be mapped into disjoint subsets of S
via the accessible region. (See Fig. 12.4.)

Finally, one could demand even less. For simplicity let f and g be real-valued
functions on X_O and X_S, respectively, and map out the graph composed of the pair
of values (o, s) where $o = f(x_O)$, $s = g(x_S)$, and x runs through all points in the
accessible region A. Suppose that the resulting graph has a shape which allows for
statistical inference from (sets of) o-values to (sets of) s-values. To the extent this
statistical inference is reliable it is possible to associate (sets of) microstates of the
observer with (sets of) microstates of the system, and thus in this weak sense we

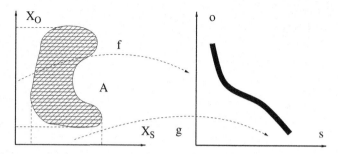

Fig. 12.5 Inverse functions f^{-1} and g^{-1} "almost" partition the phase spaces X_O and X_S

could speak of the observer being able to discern different macrostates. In case the graph is a graph of a one-to-one function we get back our first, strict definition of a macrostate, but it is easy to see that there could be other physical situations (such as the one depicted on Fig. 12.5) which could still warrant relying on this weaker concept.

In general, if one follows the path charted out by Hemmo and Shenker, the macrostate concept needs to be tailored to physical characteristics of the measurement procedure. In cases when the joint system needs to be in an equilibrium microstate in order for the measurement to count as such—in fact this is the case in the temperature measurement example above—, one would either need to apply one of the weaker notions of a macrostate, or apply the partition-to-partition macrostate notion but impose further restrictions on the accessible region A in order to exclude non-equilibrium microstates. The weakest macrostate notion may be warranted when, with an even more emphasized epistemic bent, one takes into account measurement errors, noises, environmental effects, fuzziness of macroscopic concepts, and so on.

In what follows we focus on the notion of macrostate only in the first, strongest sense, namely when *partitions* are transformed into one another in a one-to-one way. In the following two sections we will make the notion of macrostate mathematically precise and return to their physical interpretation in Sect. 12.5.

12.3 Macrostates: The Formal Definition

Consider a joint system $O + S$ with accessible region A. Denote by $2^A, 2^{A_O}$ and 2^{A_S} the *power set* of the accessible region A, and the *power set* of its projections A_O and A_S, respectively. The accessible region A will uniquely define the following maps:

$$\mathcal{A}^o : A_O \to 2^A; \quad \mathcal{A}^o(x_O) := \{x' \in A \mid x'_O = x_O\} \tag{12.1}$$

$$\mathcal{A}^s : A_S \to 2^A; \quad \mathcal{A}^s(x_S) := \{x' \in A \mid x'_S = x_S\} \tag{12.2}$$

$$\mathcal{A}^{os} : A_O \rightarrow 2^{A_S}; \quad \mathcal{A}^{os}(x_O) := \left(\mathcal{A}^o(x_O)\right)_S \tag{12.3}$$

$$\mathcal{A}^{so} : A_S \rightarrow 2^{A_O}; \quad \mathcal{A}^{so}(x_S) := \left(\mathcal{A}^s(x_S)\right)_O \tag{12.4}$$

Intuitively, $\mathcal{A}^o(x_O)$ is picking out those microstates of the joint system which are in A and the projection of which onto X_O is x_O. $\mathcal{A}^{os}$ is simply the composition of $\mathcal{A}^o$ and the projection onto X_S; it maps elements of A_O onto subsets of A_S. The functions $\mathcal{A}^s$ and $\mathcal{A}^{so}$ are defined similarly.

One can also "lift up" the above maps to the level of power sets:

$$\mathcal{A}^o[\ ] : 2^{A_O} \rightarrow 2^A; \quad \mathcal{A}^o[B_O] := \left\{x' \in A \mid x'_O \in B_O\right\} \tag{12.5}$$

$$\mathcal{A}^s[\ ] : 2^{A_S} \rightarrow 2^A; \quad \mathcal{A}^s[B_S] := \left\{x' \in A \mid x'_S \in B_S\right\} \tag{12.6}$$

$$\mathcal{A}^{os}[\ ] : 2^{A_O} \rightarrow 2^{A_S}; \quad \mathcal{A}^{os}[B_O] := \left(\mathcal{A}^o[B_O]\right)_S \tag{12.7}$$

$$\mathcal{A}^{so}[\ ] : 2^{A_S} \rightarrow 2^{A_O}; \quad \mathcal{A}^{so}[B_S] := \left(\mathcal{A}^s[B_S]\right)_O \tag{12.8}$$

Let $\mathbf{P}^A$, $\mathbf{P}^{A_O}$ and $\mathbf{P}^{A_S}$ denote the *set of partitions* of A, A_O and A_S, respectively. Let $P^{A_O} \in \mathbf{P}^{A_O}$ and $P^{A_S} \in \mathbf{P}^{A_S}$ be two partitions. Note that $\mathbf{P}^{A_O} \neq \mathbf{P}^A_O$, that is, the partitions of A_O are *not* the projections of the partitions of A. We will apply the maps (12.5)–(12.8) also to the partitions:

$$\mathcal{A}^{os}[P^{A_O}] := \{\mathcal{A}^{os}[B_O] \mid B_O \in P^{A_O}\}$$

Note that $\mathcal{A}^{os}[P^{A_O}]$ is not necessarily a partition of A_S.

With these notations in hand we can now define the notion of macrostates introduced in the previous section:

Definition 1 Let $\{P^{A_O}, P^{A_S}\}$ be a pair of partitions such that $\mathcal{A}^{os}[P^{A_O}] = P^{A_S}$ and $\mathcal{A}^{so}[P^{A_S}] = P^{A_O}$. Then we call an element of P^{A_S} a *macrostate* of the system S relative to O with accessible region A.

Next we provide a characterization of macrostates which is equivalent to the above definition. To this aim we introduce the notion of projective connectedness.

Definition 2 Let x and x' be microstates in A. We call x and x' *projectively connectible* in A, $x \sim x'$, if there exists a finite sequence $\{x_n\}_{n=1}^N$ with all $x_n \in A$ such that $x_1 = x$, $x_N = x'$ and for any x_n either $(x_n)_O = (x_{n+1})_O$ or $(x_n)_S = (x_{n+1})_S$. In other words, $x \sim x'$ iff x and x' can be connected by moving within A only along projections onto X_O and X_S, respectively. A region B in A is called *projectively connectible* if $x \sim x'$ for any $x, x' \in B$. (See Fig. 12.6.) A set $\{B_i\}$ of regions in A is called *mutually projectively unconnectible* if $x_i \not\sim x_j$ for any $x_i \in B_i$ and $x_j \in B_j$ with $i \neq j$.[1]

---

[1] Our definition of projectively connectible points allow only finite number of steps of moving along projections. In case the phase space has a topological structure the definition could be

Fig. 12.6 A projectively
connectible region

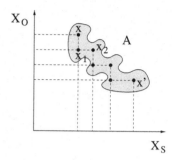

Denote a partition of A by $P^{A\sim}$ if its elements are mutually projectively unconnectible. Denote the set of such partitions of A by $\mathbf{P}^{A\sim}$. Note that $P^{A\sim}_{max}$, the partition generated by the equivalence classes of the equivalence relation $\sim$ is only one (the finest) partition of all partitions $P^{A\sim}$. A partition P is called *finer* than a partition P' if for any $B' \in P'$ there exists a set $\{B_i\}$ of elements in P such that $B' = \cup_i B_i$. P is *strictly finer* than P' if at least one set $\{B_i\}$ contains at least two elements.

Let $P^{A\sim}_O$ and $P^{A\sim}_S$ denote the projections of $P^{A\sim}$ onto X_O and X_S, respectively. Now, we formulate a new definition of macrostates.

Definition 3 Let $P^{A\sim}$ be a mutually projectively unconnectible partition. Then we call an element of $P^{A\sim}_S$ an *unconnectibility macrostate* of the system S relative to O with accessible region A.

However, as the following Proposition shows, Definitions 1 and 3 of macrostates are equivalent. Therefore we omit the adjective "unconnectibility."

Proposition 1 *A subset of A_S is a macrostate iff it is an unconnectibility macrostate.*

Proof For the proof of Proposition 1 see the Appendix.

12.4   Macrostates are Observer-Relative

Macrostates are observer-relative. If instead of O another observer, O', interacts with the system then the partition of the system's phase space induced by the interaction can be different. Measuring a container of gas by a thermometer and by a manometer do not partition the phase space of the gas in a same way.

---

straightforwardly extended to allow for the points to be projectively connected in the infinite limit. Such extension of the definition would not alter the result of Propositions 1 and 1.

Observer-dependence can be understood in two different ways. Let O and O' be two different observers who *separately* perform measurements on the same system S. Let X_{O+S} and $X_{O'+S}$ denote the phase space of the joint system $O + S$ and $O' + S$, respectively. Now, due to the different nature of interactions between the system and the different observers the accessible region A_{O+S} of the joint system $O + S$ can be different from the accessible region $A_{O'+S}$ of the joint system $O' + S$. Since macrostates of the system S are generated by the accessible region of the joint system, there is no a priori guarantee that the macrostates of S relative to $O + S$ and $O' + S$ will be the same.

Observer-dependence can also be understood, however, in a different way. Let $O+O'+S$ be the joint system of the *two* observers who perform a *joint* measurement on the system S. This is the case when we perform the temperature measurement and the pressure measurement simultaneously. Let $X_{O+O'+S}$ denote the phase space and let A denote the accessible region of the joint system. Let the phase spaces and the microstates of the subsystems be denoted just as above.

What are the macrostates of the triply joint system $O + O' + S$? One can proceed here either in a permissive or in a restrictive way. The permissive characterization of the macrostates is to say that there is a partition P^{A_O} of A_O, a second partition $P^{A_{O'}}$ of $A_{O'}$ and two *more* partitions P^{A_S} and P'^{A_S} of A_S such that the (common) accessible region A of the joint system $O + O' + S$ sends the partition P^{A_O} into P^{A_S}, and the partition $P^{A_{O'}}$ into P'^{A_S} in a one-to-one way. Hence we obtain the following definition:

Definition 1 Let $\{P^{A_O}, P^{A_S}\}$ and $\{P^{A_{O'}}, P'^{A_S}\}$ be two pairs of partitions of A_O, $A_{O'}$ and A_S, respectively, such that

$$\mathcal{A}^{os}[P^{A_O}] = P^{A_S} \quad \mathcal{A}^{so}[P^{A_S}] = P^{A_O} \tag{12.9}$$

$$\mathcal{A}^{o's}[P^{A_{O'}}] = P'^{A_S} \quad \mathcal{A}^{so'}[P'^{A_S}] = P^{A_{O'}} \tag{12.10}$$

Then we call an element of P^{A_S} and P'^{A_S} a *macrostate* of the system S relative to O and O', respectively, within the joint system $O + O' + S$ with accessible region A.

The restrictive characterization would be, however, to demand that both observers generate the *same partition* of A_S, that is to demand that there is only *one common set of macrostates* of S associated to both observers:

Definition 2 Let $\{P^{A_O}, P^{A_{O'}}, P^{A_S}\}$ be a triple of partitions of A_O, $A_{O'}$ and A_S, respectively, such that

$$\mathcal{A}^{os}[P^{A_O}] = P^{A_S} \quad \mathcal{A}^{so}[P^{A_S}] = P^{A_O} \tag{12.11}$$

$$\mathcal{A}^{o's}[P^{A_{O'}}] = P^{A_S} \quad \mathcal{A}^{so'}[P^{A_S}] = P^{A_{O'}} \tag{12.12}$$

Then we call an element of P^{A_S} a *common macrostate* of the system S relative to both O and O' within the joint system $O + O' + S$ with accessible region A.

Fig. 12.7 An accessible region with no (non-trivial) partitions $\{P^{A_{O'}}, P^{A_S}\}$

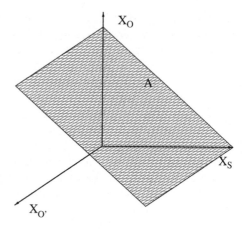

Obviously, whether two observers generate one common partition, two different partitions, or no partition at all essentially depends on the accessible region A of the joint system $O + O' + S$. In what follows we give examples for all three cases:

Example 1 A joint system $O + O' + S$ with no (non-trivial) macrostates.
   Let A be such that for any $x, y \in A$

   (i) $\mathcal{A}^{so}[X_S] = X_O$;
   (ii) $\mathcal{A}^{so}(x_S) \neq \mathcal{A}^{so}(y_S)$ if $x_S \neq y_S$;
   (iii) $\mathcal{A}^{so'}[\{x_S\}] = X_{O'}$.

(See Fig. 12.7.)
   In this case there are no (non-trivial) partitions $\{P^{A_{O'}}, P^{A_S}\}$ which would satisfy (12.10).

Example 2 A joint system $O + O' + S$ with separate partitions $\{P^{A_O}, P^{A_S}\}$ and $\{P^{A_{O'}}, P'^{A_S}\}$, but without common partition $\{P^{A_O}, P^{A_{O'}}, P^{A_S}\}$.
   Suppose that the accessible region A of the joint system is discrete and consists of only three points:

$$a = (x_1, y_1, z_1)$$
$$b = (x_1, y_2, z_2)$$
$$c = (x_2, y_2, z_3)$$

where the coordinates x, y and z refer to the microstates in X_O, $X_{O'}$ and X_S, respectively. Then the separate partitions are the following:

$$P^{A_O} = \{\{x_1\}, \{x_2\}\} \quad P^{A_S} = \{\{z_1, z_2\}, \{z_3\}\}$$
$$P^{A_{O'}} = \{\{y_1\}, \{y_2\}\} \quad P'^{A_S} = \{\{z_1\}, \{z_2, z_3\}\}$$

but it is easy to see that there is no (non-trivial) common partition.

Example 3 A joint system $O + O' + S$ with a common partition $\{P^{A_O}, P^{A_{O'}}, P^{A_S}\}$.
Modify the above example as follows. Let the three microstate be the following:

$$a = (x_1, y_1, z_1)$$
$$b = (x_2, y_2, z_1)$$
$$c = (x_3, y_3, z_2)$$

Then the common partition is the following:

$$P^{A_O} = \{\{x_1, x_2\}, \{x_3\}\} \quad P^{A_S} = \{\{z_1\}, \{z_2\}\}$$
$$P^{A_{O'}} = \{\{y_1, y_2\}, \{y_3\}\} \quad P'^{A_S} = \{\{z_1\}, \{z_2\}\}$$

To characterize the different possibilities we analyse Definitions 1 and 2 in terms of projective connectibility.

Consider the projection A_{OS} of the accessible region A of the joint system $O + O' + S$ onto X_{OS}. Define the equivalence relation $\sim$ on A_{OS} in the spirit of Definition 2. Let $P^{A_{OS}\sim}$ denote a partition of A_{OS} if its elements are mutually projectively unconnectible in the sense defined in Sect. 12.2: for any $x \in B$ and $x' \in B'$ where B and B' are different elements of $P^{A_{OS}\sim}$, $x \not\sim x'$. Denote the set of such partitions of A_{OS} by $\mathbf{P}^{A_{OS}\sim}$. Finally, denote by $P_O^{A_{OS}\sim}$ and $P_S^{A_{OS}\sim}$ the projections of $P^{A_{OS}\sim}$ onto A_O and A_S, respectively. Let $P^{A_{O'S}\sim}$, $P_{O'}^{A_{O'S}\sim}$ and $P_S^{A_{O'S}\sim}$ be defined similarly.

This leads to the following two definitions of macrostates which, however, are equivalent to the previous ones.

Definition 3 Let $P^{A_{OS}\sim}$ and $P^{A_{O'S}\sim}$ be partitions of A_{OS} and $A_{O'S}$, respectively. Then we call an element of $P_S^{A_{OS}\sim}$ and $P_S^{A_{O'S}\sim}$ an *unconnectibility macrostate* of the system S relative to O and O', respectively, within the joint system $O+O'+S$ with accessible region A.

Definition 4 Let $P^{A_{OS}\sim}$ and $P^{A_{O'S}\sim}$ be partitions of A_{OS} and $A_{O'S}$, respectively, such that $P_S^{A_{OS}\sim} = P_S^{A_{O'S}\sim}$. Then we call an element of $P_S^{A_{OS}\sim}$ an *unconnectibility common macrostate* of the system S relative to both O and O' within the joint system $O + O' + S$ with accessible region A.

Proposition 1 *Definition 1 of macrostates and Definition 3 of unconnectibility macrostates are equivalent. Definition 2 of common macrostates and Definition 4 of unconnectibility common macrostates are equivalent.*

The proof is a straightforward consequence of the proof of Proposition 1, therefore we omit it. We simply illustrate it on Examples 2 and 3 above.

In Example 2:

$$P^{A_{OS}\sim} = \{\{(x_1, z_1), (x_1, z_2)\}, \{(x_2, z_3)\}\}$$
$$P^{A_{O'S}\sim} = \{\{(y_1, z_1)\}, \{(y_2, z_2), (y_2, z_3)\}\}$$

In Example 3:

$$P^{A_{os}\sim} = \{\{(x_1, z_1), (x_2, z_1)\}, \{(x_3, z_2)\}\}$$

$$P^{A_{o's}\sim} = \{\{(y_1, z_1), (y_2, z_1)\}, \{(y_3, z_2)\}\}$$

How do two observers take notice that their joint measurement generates a common macrostate on the system S? Since in case of a common partition $\mathcal{A}^{os}[P^{A_o}] = P^{A_S} = \mathcal{A}^{o's}[P^{A_{o'}}]$, therefore whenever the microstate of O is in $B_O \in P^{A_o}$, the microstate of O' will be in $(\mathcal{A}^{so'} \circ \mathcal{A}^{os})(B_O) \in P^{A_{o'}}$. That is elements of P^{A_o} and $P^{A_{o'}}$ will be *perfectly correlated*. Thus, a system of common macrostates establishes a correlation between the measurement outcomes of the different observers.

Can the situation be reversed? Suppose the measurement outcomes of two observers O and O' of a system S are perfectly correlated in the sense that whenever the microstate of O is in $B_O \in P^{A_o}$, the microstate of O' will be in a particular $B_{O'} \in P^{A_{o'}}$. Does this perfect correlation guarantee that there is a common macrostate of the system S relative to O and O', respectively? The next example shows that even less is false.

Example 4 Perfect correlation between the observers does not even guarantee that there are separate macrostates of the joint system.

Suppose again that the accessible region A of the joint system is discrete and consists of the following four points:

$$a = (x_1, y_1, z_1)$$
$$b = (x_1, y_1, z_2)$$
$$c = (x_2, y_2, z_1)$$
$$d = (x_2, y_2, z_2)$$

where again the coordinates x, y and z refer to the microstates of O, O' and S, respectively. Here the partitions

$$P^{A_o} = \{\{x_1\}, \{x_2\}\}$$
$$P^{A_{o'}} = \{\{y_1\}, \{y_2\}\}$$

will perfectly correlate, but there is neither a $\{P^{A_o}, P^{A_S}\}$ nor a $\{P^{A_{o'}}, P^{A_S}\}$ (nontrivial) partition satisfying (12.9)–(12.10). Thus, perfect correlation between the measurements of the two observers does not guarantee a perfect correlation between any of the observers and the system.

Due to the absence of any correlation between the observers and the system in the above example one is not even justified to call O and O' observers of the system. It is straightforward to see however that if O is indeed an observer, that is O is perfectly correlated with S and also with O', then O' will also do so with S: if $\{P^{A_o}, P^{A_S}\}$ is a pair of partitions satisfying (12.9) and P^{A_O} and $P^{A_{O'}}$ are perfectly correlated, then $\{P^{A_{O'}}, P^{A_S}\}$ is a pair of partitions satisfying (12.10). In short, $\{P^{A_O}, P^{A_{O'}}, P^{A_S}\}$ is a common partition of the joint system.

12.5 Distinguishability, Regularity and Measurement

Consider again the example of the temperature measurement. Suppose you measure the temperature of a container of gas with a thermometer. Denote the thermometer by O and the gas by S. By inserting the thermometer into the gas the experimenter introduces an interaction between the measuring device and the system. This interaction fixes which microstates are possible for the joint system $O + S$, that is it fixes the accessible region A. The crucial point in the coming about of macrostates is the "shape" of the accessible region. We have seen that A can theoretically be of such shape that for any microstate of O there is only one possible microstate of S. Typically, however, the accessible region is such that only *sets* of microstates of the phase spaces X_O and X_S transform mutually into one another. That is certain sets of microstates of the thermometer will correlate with certain sets of microstates of the gas in a one-to-one manner.

But how does this abstract schema relate to the real-world temperature measurement? Measuring the temperature colloquially means the reading off of the mercury level of the thermometer inserted into the gas. The different mercury levels will again generate a partition of X_O since each microstate of the thermometer uniquely determines a mercury level. But how is this partition of X_O generated by the mercury levels is related to the above partition of X_O standing in a one-to-one correlation with a partition of X_S? This question leads us to the notion of distinguishability.

So far we have been somewhat vague about what we take to be an observer. By the same symbol O we referred to a human observer as well as to a measuring device. The reason for this ambiguity is that a direct observation of the thermometer by the sense-organs of a human observer and the measurement of the gas with a thermometer can be described in similar terms; namely, in terms of a correlation between an "observer" and a system. And this is the point where distinguishability appears. Distinguishability is nothing but an observer-system correlation such that the observer is a human observer.

Suppose an experimenter is reading off the temperature from a thermometer. The mercury levels in the glass tube are distinguishable to the naked eye while the positions of mercury molecules are not. What does it mean that the experimenter is able to distinguish between the different mercury levels of the thermometer? Physically speaking looking at the thermometer is an interaction between the thermometer and the observer's sense-organs and brain. Although the details of this interaction are highly complex, one can still think of this interaction as a constraint on the "phase space" of the joint system composed of the observer's sense-organs and brain and the thermometer. One can then speculate that the clustering of the microstates of the thermometer into distinguishable macrostates is nothing but a one-to-one correlation between certain sets of microstates of the observer's sense-organs and brain and certain other sets of microstates of the thermometer. Being able to distinguish then physically means that certain clusterings of the state space of a physical object strongly correlate with certain clusterings of our sense organ and brain states. If this speculation is correct, then it may well be the case that the accessible region of the joint system is responsible for certain stable clusterings of our brain states and for the bringing about of the associated mental state. The short cognitive story would then be this: mental states arise where the brain can make distinctions; and making distinctions is a one-to-one correlation between coarse-grained brain states and coarse-grained states of the outer world.

According to the above description both the temperature of a gas and the lengths of the mercury column, as macrostates of the corresponding systems, are observer-relative notions. The former is defined in terms of the interaction with the thermometer, the latter is defined in terms of the interaction with the human observer. Notice that since these two kinds of macrostates refer to different interactions with different "observers", there is no a priori guarantee that the two partitions of the thermometer, namely the one generated by the distinguishable mercury levels and the one standing in a one-to-one correlation with the temperature partition of the gas, coincide. It is a contingent fact of the world that the "mercury level" reflects the temperature of the gas and at the same is time accessible for human observation.

Despite the essential observer-relativity of macrostates, there is a sense in which they are inherent to the target system in question. This sense is provided by the fact that macrostates, as subsets and partitions of the system's phase space, satisfy regularities. Consider the example of the ideal gas law. The temperature T, the pressure P and the volume V are functions on phase space. Their inverse images carve out three distinct partitions of the system's phase space corresponding to the level surfaces of these phase functions. The elements of these partitions are macrostates since they correlate with elements of the phase space partition of the appropriate measuring devices via the measurement interaction. Nevertheless, there is something inherent in the partitions generated by phase function T, P and V: they stand in a well-defined functional relationship to one another *independently of whether they are measured or not*. Namely, if the ideal gas is in a microstate in which the temperature is T and the pressure is P then its volume V is proportional to $1/PT$. Again, these kinds of regularities are present regardless of being observed or not. They can be discovered, however, by an observer if she has empirical access to these partitions by various measurements. Empirical access means that the sets of partitions featuring in thermodynamic regularities and the sets of partitions brought about by the measurements coincide or are close enough to one another. But the coincidence of the two partitions, the ones brought about by measurement and the ones featuring in thermodynamic regularities, is a contingent fact of nature, as Hemmo and Shenker rightly stress. It is contingent in the sense that the two types of partitions are determined by different physical conditions: the characteristics of the interaction between the measurement device and the system on the one hand and the characteristics of the interactions within the system on the other. There is no a priori connection between these two types of conditions. One is left with this kind of contingency even if one *defines* "measurement" as a physical process revealing the system's inherent properties featuring in thermodynamic regularities. For even if the temperature measurement is *defined* as a process carving out the very partitions generated by the phase function T, it still remains a contingent fact of our world whether such physical processes do exist, and if so, what they are. One can well imagine a possible world where the functional relationship of phase functions T, P and V, as inherent properties of a system, are the same as in the actual one but the physical principles on which experiencing and measuring temperature are based (thermal expansion, heat transfer, equalization of temperature, blackbody radiation, temperature dependence of electrical resistance, etc.) are different or do not exist at all. In a possible world where these principles do not exist, temperature—despite featuring in thermodynamic regularities—wouldn't qualify as a macrostate in the sense discussed in this paper.

It is worth noting that, despite their conceptual difference, the emergence of partitions brought about by interaction with an observer and by thermodynamic regularities can be accounted for in the same mathematical framework. At the end of Sect. 12.2 (see Fig. 12.5) we defined macrostates via functions on the phase space of the system and the observer: we considered two functions, an f on X_O and a g on X_S, such that the value of one function allowed for statistical inference for the value of the other. Then we defined macrostates by the total inverse of the function f. This procedure, however, can also be applied when (i) we have more than two functions and (ii) these functions are all defined on X_S. Again, if there is a statistical relationship among the values of these phase space functions, then each function will define a macrostate via the total inverse. But this is exactly how the phase space functions carve up the phase space in thermodynamic regularities. Thus, the core idea behind bringing about partitions via interaction with an observer and by thermodynamic regularities is the same: it is a *correlation*, either in the joint phase space of the system and the observer or within the system's own phase space.

Finally, let us make a general observation on the observer-relativeness of macrostates investigated formally in the previous section. Suppose that we measure the temperature of a container of gas with a mercury thermometer O and by an alcohol thermometer O'. Measuring the temperature with a mercury thermometer generates a partition of X_S and thus brings about macrostates relative to O, as explicated earlier. Measuring the temperature with an alcohol thermometer brings about macrostates relative to O'. There is no a priori guarantee that these two partitions have anything in common. However, in practice we observe a one-to-one correlation between the readings of the mercury thermometer and the alcohol thermometer. As we saw in the end of the previous section, whenever O and O' are perfectly correlated, and O and O' are observers of a system S in the sense of there being correlation between O and S and between O' and S, then there always exits a system of common macrostates relative to O and O'. This common macrostate system can be regarded as a common causal explanation of the perfect correlation of measurements O and O'. It is the temperature of the gas, an inherent property of the system pertaining to a partition of its phase space, that explains the correlation of the two kinds of "temperature" measurement.

12.6   Conclusions

In this paper we defined macrostates, following Hemmo and Shenker, as elements of certain partitions of the system's phase space generated by the interaction between the system and an observer. We investigated several formal and conceptual features of this notion and proved the equivalence of two different characterizations of macrostates. Finally, we intended to accommodate the other two features of macrostates, namely thermodynamic regularity and distinguishability within this framework.

Acknowledgements This work has been supported by the Hungarian Scientific Research Fund, OTKA K-115593 and by the Bilateral Mobility Grant of the Hungarian and Polish Academies of Sciences, NM-104/2014.

Appendix

Proof of Proposition 1 We prove Proposition 1 via proving four lemmas.

Lemma 1 $P_O^{A\sim}$ and $P_S^{A\sim}$ are partitions of A_O and A_S, respectively, that is $P_O^{A\sim} \in \mathbf{P}^{A_O}$ and $P_S^{A\sim} \in \mathbf{P}^{A_S}$.

Proof We show that the sets of $P_O^{A\sim}$ are disjoint and add up to A_O. (For $P_S^{A\sim}$ the proof is similar.)

Suppose that sets of $P_O^{A\sim}$ are *not* disjoint, that is there exists a $b \in (B_O \cap B_O')$ with B and B' being different elements in $P^{A\sim}$. Then there exist an $x \in B$ and an $x' \in B'$ such that $x_O = x_O' = b$. But then $x \sim x'$. Contradiction.

Suppose that sets of $P_O^{A\sim}$ are *not* adding up to A_O, that is there exists a $b \in A_O$ such that $b \notin B_O$ for any $B_O \in P_O^{A\sim}$. Then $\mathcal{A}^o(b) \cap B = \emptyset$ for any $B \in P^{A\sim}$, that is $P^{A\sim}$ is not a partition of A. Contradiction. ∎

Lemma 2 *The partition $P^{A\sim}$ can be "reconstructed" from its projections in the sense that $\mathcal{A}^o[P_O^{A\sim}] = \mathcal{A}^s[P_S^{A\sim}] = P^{A\sim}$.*

Proof We show that $\mathcal{A}^o[P_O^{A\sim}] = P^{A\sim}$. (For $\mathcal{A}^s[P_S^{A\sim}] = P^{A\sim}$ the proof is similar.)

Suppose to the contrary that $\mathcal{A}^o[P_O^{A\sim}] \neq P^{A\sim}$. This means that there exist an $x \in A$ and a $B \in P^{A\sim}$ such that either (i) $x \in \mathcal{A}^o[B_O]$ and $x \notin B$, or (ii) $x \notin \mathcal{A}^o[B_O]$ and $x \in B$.

As for case (i), since $x \in \mathcal{A}^o[B_O] \setminus B$ and therefore $x_O \in B_O$, there exists an $x' \in B$ such that $x_O' = x_O$. But then $x \sim x'$ and hence $x \in B$. Contradiction.

Case (ii) can be excluded since for any region B in A, $B \subseteq \mathcal{A}^o[B_O]$. ∎

The following Lemma is a straightforward corollary of Lemma 2.

Lemma 3 $\mathcal{A}^{os}[P_O^{A\sim}] = P_S^{A\sim}$ and $\mathcal{A}^{so}[P_S^{A\sim}] = P_O^{A\sim}$.

Proof $\mathcal{A}^{os}[P_O^{A\sim}] = \left(\mathcal{A}^o[P_O^{A\sim}]\right)_S = \left(P^{A\sim}\right)_S = P_S^{A\sim}$ and similarly for $\mathcal{A}^{so}[P_S^{A\sim}]$. ∎

Obviously, Lemma 3 shows that macrostates according to Definition 3 are macrostates also according to Definition 1. Our next Lemma demonstrates that the converse is also true.

Lemma 4 *Suppose that $\{P^{A_O}, P^{A_S}\}$ are partitions of A_O and A_S, respectively, such that $\mathcal{A}^{os}[P^{A_O}] = P^{A_S}$ and $\mathcal{A}^{so}[P^{A_S}] = P^{A_O}$. Then there exists a $P^{A\sim} \in \mathbf{P}^{A\sim}$ such that $P^{A_O} = P_O^{A\sim}$ and $P^{A_S} = P_S^{A\sim}$.*

Proof (i) First we show that $\mathcal{A}^o[P^{A_O}]$ is a partition of A, that is $\mathcal{A}^o[P^{A_O}] \in \mathbf{P}^A$.

Suppose that sets of $\mathcal{A}^o[P^{A_O}]$ are *not* disjoint, that is for a given $B_O, B_O' \in P^{A_O}$ there exists an x such that $x \in (\mathcal{A}^o[B_O] \cap \mathcal{A}^o[B_O'])$. But then $x_O \in \left(B_O \cap B_O'\right)$, that is P^{A_O} is not a partition of A_O. Contradiction.

Suppose that sets of $\mathcal{A}^o[P^{A_O}]$ are *not* adding up to A, that is there exists an x in A such that $x \notin B$ for any $B \in \mathcal{A}^o[P^{A_O}]$. Then $x_O \notin B_O$ for any $B_O \in P^{A_O}$, that is P^{A_O} is again not a partition of A_O. Contradiction.

Hence, $\mathcal{A}^o[P^{A_O}]$ is a partition of A. Similarly, $\mathcal{A}^s[P^{A_S}]$ is a partition of A.

(ii) Next we show that $\mathcal{A}^o[P^{A_O}] = \mathcal{A}^s[P^{A_S}]$.

Let $B_O \in P^{A_O}$ and let B_S be its corresponding element $B_S = \mathcal{A}^{os}[B_O]$. Suppose that $\exists x \in \mathcal{A}^s[B_S]$ such that $x \notin \mathcal{A}^o[B_O]$. Then $x_O \notin B_O$ and hence $x_O \in B_O'$ for some $B_O' \neq B_O$, $B_O' \in P^{A_O}$. But since $\mathcal{A}^{os}[B_O'] = B_S'$ for some $B_S' \neq B_S$, $B_S' \in P^{A_S}$, we would have $x_S \in B_S'$ and thus $x \in \mathcal{A}^s[B_S']$ where $\mathcal{A}^s[B_S'] \neq \mathcal{A}^s[B_S]$, a contradiction. The argument is similar for the case when $\exists x \in \mathcal{A}^o[B_O]$ such that $x \notin \mathcal{A}^s[B_S]$. Hence for all $B_O \in P^{A_O}$ and for their corresponding $B_S \in \mathcal{A}^{os}[B_O]$ we have $\mathcal{A}^o[B_O] = \mathcal{A}^s[B_S]$, and thus $\mathcal{A}^o[P^{A_O}] = \mathcal{A}^s[P^{A_S}]$.

(iii) Finally, we show that $\mathcal{A}^o[P^{A_O}] = \mathcal{A}^s[P^{A_S}] \in \mathbf{P}^{A\sim}$.

Suppose that $\mathcal{A}^o[P^{A_O}] \notin \mathbf{P}^{A\sim}$, that is there exists an $x \in B$ and an $x' \in B'$ such that $x \sim x'$ and B and B' are distinct elements of $\mathcal{A}^o[P^{A_O}]$. Let $\{x_n\}_{n=1}^N$ be the sequence connecting x and x' and suppose that $(x_2)_S = x_S$ (the argument is similar when $(x_2)_O = x_O$). Now, x_2 cannot be in a $B_2 \in \mathcal{A}^o[P^{A_O}]$ such that $B_2 \neq B$, otherwise $(x_2)_S$ were in $B_S \cap (B_2)_S$ and consequently B_S and $(B_2)_S$ were not disjoint. So $x_2 \in B$. By induction, we obtain that $x_3, \ldots x_N = x'$ are all in B. Hence B and B' are not distinct. Contradiction. ∎

By this we also complete the proof of Proposition 1. ∎

Reference

Hemmo, Meir, and Orly Shenker. 2012. *The Road to Maxwell's Demon: Conceptual Foundations of Statistical Mechanics*. Cambridge: Cambridge University Press.

Chapter 13
Experimental Logics as a Model of Development of Deductive Science and Computational Properties of Undecidable Sentences

Michał Tomasz Godziszewski

13.1 Knowability as Algorithmic Learnability

Emergence and development of recursion theory and computer science enable us to rigorously address the question of characterising the class of mathematical concepts that are cognitively accessible to computational devices such as human minds. The answer to this question would give reasons for which some concepts are epistemically easy (e.g. provable within first-order theories or possessing certain combinatorial properties) and the others are cognitively hard for the human mind.

Our explication of learnable concepts is based on the assumption that human mind is a computing device. What we mean by this is that functions computable by human beings are exactly Turing-computable. In fact, we assume that human mind does not exhibit any non-recursive behaviour. A short reflection should convince us that people can perform any computation, provided sufficient amount of time and space. This leads us to the second assumption that cognitively accessible world is potentially infinite. In general, computations are unbounded with respect to required resources, like time and space. The latter are provided by the actual world. We can think of the world as if it was finite. However, we can always somehow finitely extend the actual world to fulfill our computational requirements.

Suppose we want to cope with the problem $P = \{x : \exists y R(x, y)\}$, where R is recursive. We may approach any instance "$a \in P$?" in the following way. Set answer to *no*. Start generating elements from the universe. For each generated element b

The author was supported by The National Science Centre, Poland (NCN), grant number 2013/11/B/HS1/04168.

M.T. Godziszewski (✉)
Logic Department, Institute of Philosophy, University of Warsaw, Warsaw, Poland
e-mail: mtgodziszewski@gmail.com

© Springer International Publishing AG 2017
G. Hofer-Szabó, L. Wroński (eds.), *Making it Formally Explicit*, European Studies in Philosophy of Science 6, DOI 10.1007/978-3-319-55486-0_13

check whether $R(a, b)$ and if so set answer to *yes* and stop; otherwise continue. This algorithm ensures that positive answers establish with certainty. Negative answers are subjected to uncertainty – there is no guarantee that there is no witness for the existential quantifier. Nevertheless, it is still a good cognitive strategy to rely on such algorithm. Justification comes from the work of mathematicians. Axiomatic method has been successfully used since Euclid of Alexandria. Nowadays, it is well recognized that the set of theorems of an axiomatic system with recursive set of axioms is recursively enumerable. Therefore, the algorithm described above applies to the work of "determining" the set of theorems of the particular axiomatic system. Intuitively, knowledge obtained through axiomatic systems by mathematicians seems fully legitimate. The reason *why* it is fully legitimate is that finding a proof may be difficult and take a long time, but once a proof is found, the answer is recursively conclusive. Hence, it seems theoretically justified why, for instance, theorems of axiomatic number theory or axiomatic set theory are cognitively accessible.

Observe that the above algorithm for "determining" P has the property that at some unspecified time of the computation the answer may change from *no* to *yes*. In general, it is impossible to recursively predict the moment after which the answer will change. Since if it was possible, we would easily construct a decision procedure for P. It turns out we accepted as cognitively sound a method that allows one mind-change and captures Σ_1^0 sets. This clearly shows, that decidability is too narrow concept to fit our purposes.

Consider the following method for "determining" the consistency of the theory of axiomatic system with recursive set of axioms: $T := \emptyset$. Set answer to *yes*. Inside infinite loop do the following. If T contains contradiction, answer *no* and stop. Otherwise generate next proof, add proved sentence to T and continue the loop. In this situation we can eventually arrive at conclusive answer if axiomatic system is inconsistent. But if the system is consistent, we are left uncertain. Similar procedure is easily applicable to any problem of the form $\{x : \forall y R(x, y)\}$, where R is recursive.

If we accepted as cognitively sound a method that proceeds by one mind change from *no* to *yes*, then we should also accept a method that allows one mind change from *yes* to *no*. However, the sets captured by the former kind of method are Σ_1^0, whereas the sets captured by the latter kind of method are Π_1^0. Since $\Sigma_1^0 - \Pi_1^0$ and $\Pi_1^0 - \Sigma_1^0$ are both non-empty, neither of these two kinds of methods is adequate for explaining knowability. We would need something stronger to capture both these classes.

We can see, that the common property of these two kind of methods is that on every input after some finite time they level off on the right answer. Going further, it seems justified to accept as cognitively sound any method that proceeds by mind-changes and on every input stabilizes. In this way we arrive at the concept of algorithmic learnability.

Definition 1 (Algorithmic learnability (Gold 1965, 1967 and Putnam 1965))
Let $A \subseteq \mathbb{N}$. Say that A is algorithmically learnable iff there is a total computable function $g : \mathbb{N}^2 \to \{0, 1\}$ such that for all $x \in \mathbb{N}$: $\lim_{t \to \infty} g(x, t) = 1 \Leftrightarrow x \in A$ and $\lim_{n \to \infty} g(x, t) = 0 \Leftrightarrow x \notin A$.

The notion of algorithmic learnability is one of the equivalent formulations of the concept of methods proceeding by mind-changes and stabilizing on every input.

For our further investigations we need the following theorem which we leave without a proof:

Theorem 1 (Generalized Limit Lemma (Shoenfield 1959)) *Let $R \subseteq \mathbb{N}^s$. Then the following are equivalent:*

1. *R is recursive with recursively enumerable oracle.*
2. *$deg(R) \leq 0'$.*
3. *R is algorithmically learnable, i.e. there exists a recursive sequence of recursive relations $S_0, S_1, S_2, \ldots = (S_n)_{n \in \omega}$ such that $\chi_R = \lim_{n \to \infty} \chi_{S_n}$.*
4. *R is Δ_2^0.*

To obtain our result, we need to use the fact certain classes of arithmetical formulae contain their own truth definiitions.

Definition 2 (Arithmetical hierarchy of formulae) For any formula (in the language of arithmetic) φ we say that φ is:

   (i) Δ_0 ($= \Sigma_0$, $= \Pi_0$) iff every quantifier occuring in φ is bounded, where bounded quantifier is a quantifier of the form: $\forall x \leq y$ and $\exists x \leq y$,[1]
   (ii) Σ_{n+1} iff it is logically equivalent to a formula of the shape $\exists x_1, \ldots, \exists x_k \psi (x_1, \ldots, x_k)$, where $\psi(x_1, \ldots, x_k)$ is Π_n,
   (iii) Π_{n+1} iff it is logically equivalent to a formula of the shape $\forall x_1, \ldots, \forall x_k \psi (x_1, \ldots, x_k)$, where $\psi(x_1, \ldots, x_k)$ is Σ_n.

As we deal with sets and relations that bear certain place in the arithmetical hierarchy, we also provide the definition of the arithmetical hierarchy of relations:

Definition 3 (Arithmetical hierarchy of relations) Let $R \subseteq \mathbb{N}^k$ be an arithmetical relation and let $\varphi(x_1, \ldots, x_k)$ be an arithmetical formula. We say that R is (arithmetically) defined by φ (or that φ defines R), iff for any $a_1, \ldots, a_k \in \mathbb{N}$ it holds that

$$(a_1, \ldots a_k) \in R \quad \Leftrightarrow \quad \mathbb{N} \models \varphi(a_1, \ldots, a_n).$$

We say that R is (arithmetically) definable if there exists an arithmetical formula φ such that φ defines R.

The relation R is Σ_n^0 iff it is definable by a Σ_n-formula, it is Π_n^0 iff it is definable by a Π_n-formula, and it is Δ_n^0 iff it is definable both by a Σ_n-formula and a Π_n-formula.

Relations that are Σ_1^0 are called recursively enumerable (or computably enumerable), and those that are Π_1^0 are called co-recursively enumerable (or co-computably enumerable).

---

[1] We obviously read $\forall x \leq y \varphi(x)$ as $\forall x(x \leq y \Rightarrow \varphi(x))$ and $\exists x \leq y \varphi(x)$ as $\exists x(x \leq y \wedge \varphi(x))$.

Theorem 2 (Definability of truth for bounded classes of sentences)

(i) *There exist formulae φ, ψ such that φ is Σ_1 and ψ is Π_1 and both φ and ψ are truth definitions for the class Δ_0 of sentences and $PA \vdash \varphi \equiv \psi$,*

(ii) *for any $n > 0$ there exists a formula Tr_{Σ_n} such that it is Σ_n and it is a truth predicate for the class Σ_n in PA,*

(iii) *for any $n > 0$ there exists a formula Tr_{Π_n} such that it is Π_n and it is a truth predicate for the class Π_n in PA.*

Proof (Sketch – we follow the proofs from R. Kaye (1991) and P. Hajek and P. Pudlak 1993.) We will first prove the part (i). First, it is easy (however tedious) to construct an arithmetical formula $ParSat_0(x, y, z)$ such that its intuitive meaning is : *x is a partial satisfaction relation for all Δ_0-formulae φ such that $\ulcorner\varphi\urcorner < y$ and for all assignments v such that for any free variable x_i in φ $v(\ulcorner x\urcorner) < z$. The statement that x is is a partial satisfaction relation is to be read as follows: x is a finite function such that its domain is a set of all pairs $(\ulcorner\varphi\urcorner, v)$, where φ is Δ_0 and $\ulcorner\varphi\urcorner \leq y$, v is an assignement for φ and all values of v are $< z$ and the range of x is a set $\{0, 1\}$.* Furthermore, we demand that the following **Tarski's conditions** hold for x:

1. If τ_1 and τ_2 are terms and if $\ulcorner\varphi\urcorner = \ulcorner\tau_1 = \tau_2\urcorner$, then $x(\ulcorner\varphi\urcorner, v) = 1$ iff $val(\tau_1, v) = val(\tau_2, v)$.
2. If $\ulcorner\varphi\urcorner = \ulcorner\psi \wedge \gamma\urcorner$, then $x(\ulcorner\varphi\urcorner, v) = 1$ iff $x(\ulcorner\psi\urcorner, v) = 1$ and $x(\ulcorner\gamma\urcorner, v) = 1$.
3. If $\ulcorner\varphi\urcorner = \ulcorner\neg\psi\urcorner$, then $x(\ulcorner\varphi\urcorner, v) = 1$ iff $x(\ulcorner\psi\urcorner, v) = 0$.
4. If $\ulcorner\varphi\urcorner = \ulcorner\exists x_i < x_j\psi(x_i, x_j)\urcorner$, then $x(\ulcorner\varphi\urcorner, v) = 1$ iff for some v_1 such that all values of v_1 are $< z$ and it differs from v at most on the variable x_i and $v_1(x_i) < v_1(x_j)$, the following holds: $x(\ulcorner\psi\urcorner, v_1) = 1$.

By $val(a, b) = c$ we understand an arithmetical formula which means that c is a value of the term a under the assignment b. We may write down the conjunction of the above conditions with a Σ_1-formula as well as with a Π_1-formula. Having written it down, we may define a full satisfaction relation for Δ_0-formulae. Let us assume we have a formula $ParSat_0(x, y, z)$ that is Σ_1.

We denote the full satisfaction predicate by $Sat_0(\ulcorner\varphi\urcorner, v)$ and define it as follows:

$$\exists x\exists y\exists z \ (ParSat_0(x, y, z) \wedge x(\ulcorner\varphi\urcorner, v) = 1)$$

As we see, the formula $Sat_0(\ulcorner\varphi\urcorner, v)$ is Σ_1. However, there is an equivalent version of $Sat_0(\ulcorner\varphi\urcorner, v)$ such that it is Π_1. We define it as follows:

$$\forall x\forall y\forall z \ ((ParSat_0(x, y, z) \wedge x(\ulcorner\varphi\urcorner, v) \text{ is defined}) \Rightarrow x(\ulcorner\varphi\urcorner, v) = 1).$$

Now we may introduce the predicate $Tr_{\Delta_0}(w)$ such that it is a truth predicate for Δ_0-sentences and it is Σ_1. We define it as follows:

$$\exists v Sat_0(\ulcorner\varphi\urcorner, v).$$

We can also introduce the predicate $Tr_{\Delta_0}(w)$ such that it is a truth predicate for Δ_0-sentences and it is Π_1. We define it as follows:

$$\forall v Sat_0(\ulcorner \varphi \urcorner, v)$$

It follows from the construction that $Tr_{\Delta_0}(w)$ is a truth definition for Δ_0-sentences.

We now turn to proving (ii) and (iii). The proof is inductive. First, we already have the truth predicate for Δ_0-sentences. Now, assume we have established a satisfaction predicate $Sat_{\Sigma_k}(\ulcorner \varphi \urcorner, v)$ for the class of Σ_k-sentences. We define $Sat_{\Pi_{k+1}}(\ulcorner \varphi \urcorner, v)$ for the class of Π_{k+1}-sentences as a conjunction of following statements:

1. $\ulcorner \varphi \urcorner = \ulcorner \forall x_1, \ldots, x_m\ \psi(x_1, \ldots, x_m) \urcorner$ for ψ being Σ_k and the tuple of variables $x_1, \ldots, x_m$,
2. v is an assignment defined for free variables occurring in φ,
3. for any assignment v_1 such that it does not differ from v on the free variables occurring in φ, $Sat_{\Sigma_k}(\ulcorner \varphi \urcorner, v)$ holds.

Such a formula is Π_{k+1}, since by the assumption $Sat_{\Sigma_k}(\ulcorner \varphi \urcorner, v)$ is Σ_k and we have the unbounded general quantifier *for any assignment v_1 such that* which makes the arithmetical rank of our formula higher.

Now, on the other hand, assume that we have established a satisfaction predicate $Sat_{\Pi_k}(\ulcorner \varphi \urcorner, v)$ for the class of Π_k-sentences. We define $Sat_{\Sigma_{k+1}}(\ulcorner \varphi \urcorner, v)$ for the class of Σ_{k+1}-sentences as a conjunction of following statements:

1. $\ulcorner \varphi \urcorner = \ulcorner \forall x_1, \ldots, x_m\ \psi(x_1, \ldots, x_m) \urcorner$ for ψ being Π_k and the tuple of variables $x_1, \ldots, x_m$,
2. v is an assignment defined for free variables occurring in φ,
3. the exists an assignment v_1 such that it does not differ from v on the free variables occurring in φ, $Sat_{\Pi_k}(\ulcorner \varphi \urcorner, v)$ holds.

Such a formula is Σ_{k+1}, since by the assumption $Sat_{\Pi_k}(\ulcorner \varphi \urcorner, v)$ is Π_k and we have the unbounded existential quantifier *there exists an assignment v_1 such that* which makes the arithmetical rank of our formula higher. $\square$

13.2 Experimental Logic

We follow here Jeroslow from 1975, whose approach diverged from the classical (dating back to Gödel) perspective on analysing computability-theoretic aspects of axiomatic systems (which obviously gave rise to great results already well-known at the time Jeroslow wrote his paper, with examples such as the ones in Craig (1953) or Feferman (1957, 1960). That is, we identify the mechanistic conception of a theory which proceeds by trial-and-error with a recursive predicate $H(t, x, y)$ of three variables interpreted intuitively as follows: At time t, the finite configuration with Gödel numer y is accepted as a justification of the formula with Gödel number x.

Definition 1 Given an experimental logic $H = H(t, x, y)$ we identify the **theorems** of H with **recurring formulae** of defined by the condition:

$$Rec_H(x) \equiv \forall t \exists s \geq t \exists y H(s, x, y).$$

We define the **stable formulae** of H by the condition:

$$Stbl_H(x) \equiv \exists t \exists y \forall s \geq t H(s, x, y).$$

We say that H is **convergent** if all recurring formulae are stable. Since the implication for the other direction is obvious by the predicate calculus, we may identify H being convergent with the following equivalence: $\forall x Rec_H(x) \equiv Stbl_H(x)$

Theorem 1 *The sets of theorems of convergent, experimental logics are precisely the Δ_2^0 sets.*

Proof By the limit lemma a set A is Δ_2^0 if and only if there exists a function $f : \omega \to \{0, 1\}$ such that:

$$\forall x (x \in A \Leftrightarrow \exists t \forall s \geq t \ f(x, s) = 1),$$

$$\forall x (x \notin A \Leftrightarrow \exists t \forall s \geq t \ f(x, s) = 0).$$

Therefore we may take $H(t, x, y) := f(x, t) = y \wedge y = 1$. Then we have $Stbl_H(x) \equiv \exists y \exists t \forall s \geq t \ (f(x, t) = y \wedge y = 1)$. $\qquad \square$

This result is crucial, since it means that theorems of convergent, experimental logic are exactly algorithmically learnable.

Our next theorem extends Gödel's incompleteness theorem (Gödel 1967) in terms of intrinsic limitations of experimental logics. From now on, by *PA*, we denote only the set of axioms of *PA* which is not to be confused with the set of logical consequences of *PA*, from now on, denoted by *Cn(PA)*.

Theorem 2 (Jeroslow 1975) *Let H be a consistent, convergent, experimental logic whose theorems contain those of first-order Peano arithmetic and whose theorems are closed under first-order predicate reasoning. Then there is a true Π_1^0 sentence that is not provable in H.*

Proof First of all, let us notice that if $\exists x \forall y \psi(x, y)$ is a true, but unprovable Σ_2^0 sentence, then for some n we have that $\forall y \psi(n, y)$ is true but unprovable Π_1^0 sentence.

By the diagonal lemma, we can easily obtain a formula φ such that:

$$\vdash Rec(\varphi) \equiv \neg \varphi. \tag{13.1}$$

We can see that φ is equivalent to a Σ_2^0 sentence. There are now two possibilities:

1. $\vdash Rec(\varphi) \Rightarrow \varphi$.
2. $\nvdash Rec(\varphi) \Rightarrow \varphi$.

Let us consider case 1 first. Since by (3) we obtained that $\vdash Rec(\varphi) \Rightarrow \neg\varphi$, by our assumption we get $\vdash \neg Rec(\varphi)$, and by (3) again we get that $\vdash \varphi$. Therefore $Stbl(\varphi)$ is a true Σ_2^0 sentence. It suffices to show that $Stbl(\varphi)$ is not provable. For the sake of contradiction, suppose $\vdash Stbl(\varphi)$. This obviously means that $\vdash Rec(\varphi)$ and from this it follow that H is inconsistent, contrary to our general assumption.

Now let us proceed with case 2. It now suffices to show that $Rec(\varphi) \Rightarrow \varphi$ is true since by its construction and assumption of our case, it is an unprovable Σ_2^0 sentence. Suppose $Rec(\varphi)$ is true. Since H is convergent, $Stbl(\varphi)$ is then true as well. Hence, we have $\vdash \varphi$. But then obviously $\vdash Rec(\varphi) \Rightarrow \varphi$, contradicting the case. Thus, $Rec(\varphi)$ is false and by trivial propositional calculus $Rec(\varphi) \Rightarrow \varphi$ is true. □

From this theorem we have an immediate, but an important corollary:

Corollary 1 *The deductive closure of $PA + \{\varphi \in \Pi_1^0 - Sent_\mathcal{L} : \mathbb{N} \models \varphi\}$ is not Δ_2^0.*[2]

Proof If such a theory was Δ_2^0, it would be a convergent experimental logic and as such it would not contain some true Π_1^0 sentence, but it does contain all of them by the definition, which would be inconsistent. □

13.3 Main Results

We are working under the assumption that the theories: PA, $PA + Con(PA)$ and $PA + \neg Con(PA)$ are consistent.

Let us define the following sets of (codes of, i.e. Gödel numbers of) arithmetical sentences:

Definition 1 1. $X := \{\ulcorner\varphi\urcorner \in \Pi_1^0 : PA + Con(PA) \vdash \varphi$ and $PA \nvdash \varphi\}$.
2. $Y := \{\ulcorner\varphi\urcorner \in \Pi_1^0 : PA \nvdash \varphi$ and $PA \nvdash \neg\varphi\}$.
3. $Z := \{\ulcorner\varphi\urcorner \in \Pi_1^0 : \mathbb{N} \models \varphi\}$.

For convenience, we will omit the corner notations from now on – the Reader is asked only to remember that while speaking of X, Y and Z, we are dealing with sets of natural numbers.

Theorem 1 $X \subset Y \subset Z$.

Proof $(X \subset Y)$
Lat us take any $\varphi \in X$. By assumption, we have $PA \nvdash \varphi$. For the sake of contradiction suppose $PA \vdash \neg\varphi$. But then, obviously $PA + Con(PA) \vdash \neg\varphi$. But

---

[2] Instead of this we can denote it more easily: $Cn(PA + \{\varphi \in \Pi_1^0 : \mathbb{N} \models \varphi\})$ is not Δ_2^0.

this means that $PA + Con(PA)$ is inconsistent, which is inconsistent with out general assumption. Now we will show that the inclusion $X \subseteq Y$ is proper. By the diagonal lemma, there is a sentence $\psi \in Sent_{\mathcal{L}}$ such that:

$$PA + Con(PA) \vdash \psi \equiv \neg Pr_{PA+Con(PA)}(\overline{\ulcorner \psi \urcorner}).$$

Oobviously $\psi \in \Pi_1^0$. Therefore by the proof of Gödel's theorem we have that $PA + Con(PA) \nvdash \psi$. Then, obviously $PA \nvdash \psi$. On the other hand ψ is true, i.e. $\mathbb{N} \models \psi$, therefore $PA \nvdash \neg\psi$. This means $\psi \in Y$ and $\psi \notin X$.

$(Y \subset Z)$

Lat us take any $\varphi \in Y$. For the sake of contradiction, suppose $\mathbb{N} \nvDash \varphi$. Then, by the definition of satisfaction (Tarskian classical semantics) $\mathbb{N} \models \neg\varphi$. However, $\neg\varphi \in \Sigma_1^0$. By Σ_1^0-completeness of PA we then obtain $PA \vdash \neg\varphi$ which is inconsistent with $\varphi \in Y$. The inclusion is proper, since every Π_1^0-sentence φ such that $PA \vdash \varphi$ is in Z, but not in Y, by the definition of both of them. $\qquad\square$

Lemma 1 $PA + \neg Con(PA) \vdash \varphi$ *is equivalent to* $PA + \neg\varphi \vdash Con(PA)$.

Proof The statement of the lemma follows easily form the following sequence of equivalent statements:

1. $PA + \neg Con(PA) \vdash \varphi$.
2. For any model $\mathcal{M}$ if $\mathcal{M} \models (PA + \neg Con(PA))$, then $\mathcal{M} \models \varphi$.
3. For any model $\mathcal{M}$ if $\mathcal{M} \nvDash \varphi$, then $\mathcal{M} \nvDash (PA + \neg Con(PA))$.
4. For any model $\mathcal{M}$ if $\mathcal{M} \models \neg\varphi$, then $\mathcal{M} \nvDash PA$ or $\mathcal{M} \models Con(PA)$).
5. For any model $\mathcal{M}$ if $\mathcal{M} \models \neg\varphi$ and $\mathcal{M} \models PA$, then $\mathcal{M} \models Con(PA)$).
6. $PA + \neg\varphi \vdash Con(PA)$.

$\qquad\square$

Lemma 2 $PA + \neg Con(PA)$ *is* Π_1^0-*conservative over PA, i.e. for any arithmetical sentence* $\varphi \in \Pi_1^0$ $PA + \neg Con(PA) \vdash \varphi$ *if and only if* $PA \vdash \varphi$.

Proof ($\Longleftarrow$) – obvious.

($\Longrightarrow$) Let us assume that $PA + \neg Con(PA) \vdash \varphi$. From the previous lemma $PA + \neg Con(PA) \vdash \varphi$ is equivalent to $PA + \neg\varphi \vdash Con(PA)$. But $\neg\varphi \in \Sigma_1^0$, and for any recursive extension of PA we have provable Σ_1^0-completeness, i.e. for any $\psi \in \Sigma_1^0$ and any T – recursive extension of PA we have: $T \vdash \psi \Rightarrow Pr_{PA}(\overline{\ulcorner \psi \urcorner})$. We therefore have:

$$PA + \neg\varphi \vdash \neg\varphi \Rightarrow Pr_{PA}(\overline{\ulcorner \neg\varphi \urcorner}).$$

But of course $PA + \neg\varphi \vdash \neg\varphi$. Hence,

$$PA + \neg\varphi \vdash Pr_{PA}(\overline{\ulcorner \neg\varphi \urcorner}).$$

This and the fact that $PA + \neg\varphi \vdash Con(PA)$ give us $PA + \neg\varphi \vdash Con(PA + \neg\varphi)$. From the second Gödel's incompleteness theorem we obtain that $\neg Con(PA + \neg\varphi)$ which is equivalent to $PA \vdash \varphi$, which ends the proof. □

Theorem 2 *The set of all Π_1^0-sentences which are unprovable in PA is many-one reducible to the set X.*

Proof Let us define an arithmetical function $f : \omega \to \omega$ such that

$$f(\ulcorner\varphi\urcorner) = \ulcorner Con(PA) \vee \varphi\urcorner.$$

We will show that

$$f(\ulcorner\varphi\urcorner) \in X \iff PA \nvdash \varphi.$$

Obviously, for any sentence φ we have $PA + Con(PA) \vdash Con(PA) \vee \varphi$. Hence, by the definition of X, $f(\ulcorner\varphi\urcorner) \in X$ if and only if $PA \nvdash Con(PA) \vee \varphi$, which is equivalent to $PA + \neg Con(PA) \nvdash \varphi$. By the previous lemma this is equivalent to $PA \nvdash \varphi$. This ends the proof. □

Corollary 2 *The set X is Π_1^0-hard.*

Proof Let $W = \{\ulcorner\varphi\urcorner \in \Pi_1^0 : PA \nvdash \varphi\}$. From the theorem above we know that $W \leq_m X$. But the set W is Π_1^0-complete – it is defined by the Π_1^0-relation, i.e.

$$\forall x \in \omega \, (x \in W \Leftrightarrow (x \in \Pi_1^0 \wedge \forall y \neg Prov(y, x))).$$

This is a Π_1^0-relation since the set of Π_1^0-sentences has its own truth definition, as we proved. It is Π_1^0-complete because its complement – the set of sentences not being Π_1^0 or provable in PA is trivially Σ_1^0-complete.[3] □

---

[3] Another way to see that X is Π_1^0-hard – explicitly using diagonalization – would be as follows (the argument below is a quotation of E. Jeřábek – a proof given in the communication via Internet, see: www.mathoverflow.net/questions/63690):

Let $\sigma(x) = \exists v \, \theta(x, v)$ be a complete Σ_1^0-formula (such that it is not equivalent to any Δ_0^0-formula, where $\theta \in \Delta_0^0$, and find a formula $\pi(x)$ such that PA proves

$$\pi(x) \equiv \forall w \, (\mathrm{Prov}_{PA}(w, \ulcorner\pi(\dot{x})\urcorner) \Rightarrow \exists v \leq w \, \theta(x, v))$$

by the diagonal lemma. Let $n \in \omega$. Since $\neg\pi(\bar{n})$ is equivalent to a Σ_1^0 sentence, PA proves $\neg\pi(\bar{n}) \Rightarrow \mathrm{Pr}_{PA}(\ulcorner\neg\pi(\bar{n})\urcorner)$. By definition, $\neg\pi(\bar{n}) \Rightarrow \mathrm{Pr}_{PA}(\ulcorner\pi(\bar{n})\urcorner)$, hence PA proves $\mathrm{Con}_{PA} \Rightarrow \pi(\bar{n})$. We claim that

$$(*) \quad \mathbb{N} \models \sigma(n) \iff PA \vdash \pi(\bar{n}),$$

which means that $n \mapsto \ulcorner\pi(\bar{n})\urcorner$ is a reduction of the Π_1^0-complete set $\{n : \mathbb{N} \models \neg\sigma(n)\}$ to X.

To show $(*)$, assume first that $\mathcal{M} \models PA + \neg\pi(\bar{n})$. Then there is no standard PA-proof of $\pi(\bar{n})$, hence the witness $w \in \mathcal{M}$ to the leading existential quantifier of $\neg\pi(\bar{n})$ must be nonstandard. Then $\neg\theta(n, v)$ holds for all $v \leq w$, and in particular, for all standard v, hence $\mathbb{N} \models \neg\sigma(\bar{n})$.

Theorem 3 $Cn(PA + Con(PA)) = Cn(PA + X)$

Proof ($\subseteq$) Let φ be such that $PA + Con(PA) \vdash \varphi$. Obviously $Con(PA) \in X$, therefore trivially $PA + X \vdash \varphi$.

($\supseteq$) Let φ be such that $PA + X \vdash \varphi$. Since this is a first-order theory, by completeness and compactness we can infer that in the proof of φ from $PA + X$ we use finitely many formulae, namely: $\phi_1, \phi_2, \ldots \phi_n$. All of them either belong to PA or belong to X or can be inferred from $PA + X$. In particular they are implied by $PA + Con(PA)$. If so, they can be used in the proof of φ form $PA + Con(PA)$, so $PA + Con(PA) \vdash \varphi$.   □

Corollary 3 *The set $Cn(PA + X)$ is Δ_2^0 (and as such: algorithmically learnable and FM-representable).*

Proof Since $(PA + Con(PA))$ is a recursive extension of PA, it is a recursively enumerable set, i.e. Σ_1^0. By the fact that it is identical with the set $Cn(PA + X)$, the latter one also must be recursively enumerable, and in particular: algorithmically learnable.   □

High complexity of X comes from excluding certain sentences – namely those sentences that are provable in PA. But adding PA and then closing under consequence restores those sentences. That is why the complexity decreases. It is not very surprising that Cn operator can decrease the complexity of a set of sentences – we can always add a negation of a sentence of any given set to obtain an inconsistent theory which will be (primitive) recursive. The above is however a very nice example of how Cn can decrease the complexity of a given theory to something higher than just a set whose characteristic function is primitive recursive.

We have shown that although the complexity of the set X of the (Gödel numbers of) Π_1^0-sentences unprovable in PA but provable in $PA + Con(PA)$ is high, the set $Cn(PA + X)$ is learnable, i.e. *easy* in terms of computational cognitive capacities. Jeroslow showed that the set $Cn(PA + Z)$ is not learnable. However, the set Z of all true Π_1^0-sentences seems to be very *big* – it contains a very large number of sentences and adjoining it to PA and closing under consequence also results in a complicated theory not very surprisingly. So a question rises: is there a way to improve Jeroslow's result by adjoining a *smaller* set to axioms of PA? The answer is YES and the set adjoined to the axioms of PA that results in a non-learnable theory after closing it under logical consequence is of particular epistemological interest – we can achieve epistemically hard, non-learnable theory by enriching PA with the set of Π_1^0-sentences undecidable in PA, namely the set: Y defined above.

On the other hand, assume that PA proves $\pi(\bar{n})$, and let k be the code of its proof. Since PA is sound, $\mathbb{N} \models \pi(\bar{n})$, hence there exists $v \leq k$ witnessing $\theta(\bar{n}, v)$, i.e. $\mathbb{N} \models \sigma(\bar{n})$, which ends the proof.

Theorem 4 $Cn(PA + Y) = Cn(PA + Z)$

Proof ($\subseteq$) Let $\varphi \in Cn(PA + Y)$. Without loss of generality, assume $PA \nvdash \varphi$. Then, in the proof of φ from $PA + Y$ there occurs a finite number of sentences that are consequences of PA and a finite number of undecidable Π_1^0-sentences. But any undecidable Π_1^0-sentence is in Z, since if it was not, it would have to be a false Π_1^0-sentence, yet its negation would be a true Σ_1^0-sentence. By Σ_1^0-completeness of PA the latter would be provable and the theory would be inconsistent, contrary to our assumption. Therefore φ is also provable from $PA + Z$, which means $\varphi \in Cn(PA + Z)$.

($\supseteq$) Let $\varphi \in Cn(PA + Z)$. Without loss of generality, assume $PA \nvdash \varphi$. Then, in the proof of φ from $PA + Y$ there occurs a finite number of sentences that are consequences of PA and a finite number of true, but unprovable Π_1^0-sentences. But such sentences are in Y, therefore φ is also provable from $PA + Y$, which means $\varphi \in Cn(PA + Y)$.   □

Corollary 4 *The set $Cn(PA + Y)$ is not Δ_2^0.*

Proof Immediate, by the fact that $Cn(PA + Z)$ is not Δ_2^0.   □

We may sum up this result in the following terms:

Corollary 5 *Undecidable sentences of arithmetical theories (recursively) extending PA are not algorithmically learnable.*

13.4 Conclusions and Final Remarks

Experimental logics framework, being in accordance with the trial-and-error learning concept, seems to be a good explication of the process of acquiring the content of (deductive) scientific concepts by the computational mind. While learning mathematical concepts, we conjecture some of its properties and search for justifications of our statements about them. If we accept some sequence of expressions as the justification for a given mathematical proposition in a given moment of time – e.g. a convincing example, it may happen that in view of new, empirical data we change our mind and abandon the justification we have. The situation in which we search for justifications of given conjectures and even sometimes adjust the notions we formalize (as it was convincingly shown by I. Lakatos in 1976) is formalized by the notion of recurring formula. Finding a correct notion, on the other hand, namely finding a proof, seems to be formalized by the notion of stable formula. Therefore, convergent logic is an idealization of a deductive apparatus such that justifications for our mathematical statements we find within the apparatus are always the proofs of those statements.

Within a computational view on mathematics presented in this paper, it is easily explainable, why some sentences in the language of our arithmetical theory are left independent of the theory and undecidable on its grounds – by the complexity of provability relations, adjoining the unprovable sentences to our arithmetics would provide us with a non-learnable theory. Such a theory would not be credible as set of epistemically accessible mathematical truths, since by the character of mathematical cognition we are not able to computationally *handle* such complicated sets.

References

Craig, William. 1953. On axiomatisability within a system. *Journal of Symbolic Logic* 18: 30–32.
Feferman, Solomon. 1957. Degrees of unsolvability associated with classes of formalized theories. *Journal of Symbolic Logic* 22: 161–175.
Feferman, Solomon. 1960. Arithmetization of metamathematics in a general setting. *Fundamenta Mathematicae* 49: 35–92.
Gödel, Kurt. 1967. On formally undecidable propositions of Principia Mathematica and related systems I. In *From Frege to Gödel. A source book in mathematical logic 1879–1931*, ed. Jean van Heijenoort, 596–616. Cambridge: Harvard University Press.
Gold, Martk E. 1965. Limiting Recursion. *Journal of Symbolic Logic* 30: 28–48.
Gold, Mark E. 1967. Language identification in the limit. *Information and Control* 10: 447–474.
Hajek, Peter and Pavel Pudlak. 1993. *Metamathematics of first-order arithmetic*. Berlin: Springer.
Jeroslow, Robert G. 1975. Experimental logics and Δ_2^0-theories. *Journal of Philosophical Logic* 4: 253–267.
Kaye, Richard. 1991. *Models of Peano arithmetic*. Oxford: Oxford University Press.
Lakatos, Imre. 1976. *Proofs and refutations*. Cambridge: Cambridge University Press.
Putnam, Hilary. 1965. Trial and error predicates and the solution to a problem of Mostowski. *Journal of Symbolic Logic* 30: 49–57.
Shoenfield, Joseph R. 1959. On degrees of unsolvability. *Annals of Mathematics* 69: 644–653.

CPSIA information can be obtained
at www.ICGtesting.com
Printed in the USA
BVOW06*1044230417

482034BV00008B/47/P

9 783319 554853